Communication Diary

Novel 'Impedance Matching' and the sequel

Kazuki Sumino

Copyright

Communication Diary
　— Novel 'Impedance Matching' and the sequel —
Kazuki Sumino

Translation by Kazuki Sumino
Cover art by Kazuki Sumino
The illustration of the person on the cover was generated by Adobe Firefly.
The source of the Smith chart on the cover is Wikipedia.
https://upload.wikimedia.org/wikipedia/commons/7/74/Smith_chart3.svg

Adobe and Firefly are registered trademarks or trademarks of Adobe in the United States and/or other countries.

This book is fiction; TELEC is a real organization, but its depiction is not current or exact. Yoshioka is modeled on a real person, but there is no precise description of that person. Other names, characters, places, and events are fictitious or figments of the author's imagination. Any similarity to actual events, places, or persons (living or dead) is coincidental.

'Communication Diary
　— Novel 'Impedance Matching' and the sequel —'

by Kazuki Sumino
Copyright © 2010-2024 Kazuki Sumino
All rights reserved.
The original Japanese edition was published by ONBOOK, Inc., Tokyo.
English translation © 2024 by Kazuki Sumino

November 20, 2010, First Japanese edition, first printing
June 1, 2016, Reprint published
November 7, 2018, Revised Kindle edition published
December 8, 2024, English Kindle edition published
Author: Kazuki Sumino

The scanning, uploading, and distributing this book without permission is a theft of the author's intellectual property. If you would like permission to use material from the book (other than for review purposes), please contact the author via email at rf.sumino@gmail.com. Thank you for your support of the author's rights.

Contents

Copyright 2
Impedance Matching 5
   Mission Impossible 6
   The truth behind the suspicion 25
   Transfer 36
   Smith chart 43
   Oscillator circuit 50
   Barrack circuit 66
   Finding the circuit values 72
   Acceptance inspection 81
   The wake 84
   Business negotiation 94
   Product development 98
   Trap 109
   Send-off party 116
50 Causes of Arriving No Messages 118
   Problem 119
   Reproduction 145
   Analysis 153
   The cause 174
   The solution 198
Glossary 209
Afterword 216
About the Author 227

Impedance Matching

Mission Impossible

"Four more poles!?"
Yuri Ueda resisted saying the rest of the words, and in her mind, she thought, "That's not what we've been talking about. The specifications have been approved. Besides, if we change it now, we won't be able to meet the deadline. I have optimized the circuit design, so we don't have that many extra ports!"
Her body heated up, and the words she swallowed beaded on her forehead in sweat. But she knew it was useless to say it. Yuri wiped the sweat from her forehead and added one more word.
"If it's two poles, I can handle it."

Yuri worked for this company for almost two years after graduating from a local, national university with a bachelor's degree in electronics and information engineering. She was slim and tall, with short-cut hair. This female engineer looked sharp, partly because she was wearing a black suit for the first time in a long time. Yuri's old-fashioned face was sharper because her dignified eyebrows were raised angrily.

It was 200X, and the setting was the suburbs of Tokyo, Japan. Starting at 10:00 a.m., the conference room of Narumi Electronics Industry Co. was the stage for a new product design review. The weak winter sunlight had become surprisingly strong in March, casting a bright light on the proceedings.

In the conference room, the blinds were closed, the wall was used as a screen, and a projector showed the materials. The meeting host was Yuri's boss, Yoshioka, a manager of Development Section 2. He was an honest, kind-looking man with tightly parted hair who had just returned from a long overseas business trip.

This product, for which Yuri designed the control circuit for the first time, was a device that collected data wirelessly. Various sensors could be connected to the peripheral unit, and the information could be sent wirelessly to the central unit. This product could be applied in multiple ways according to user needs.

After the project leader, Shibata, and the other project members had finished explaining their respective areas of responsibility, Katakura, the sales representative, said, "This morning, we received a request from the customer to change the specifications. They want us to increase the number of DIP switch settings. Four more."

Yuri responded to these words above. The design could handle two more poles. However, there were not enough input ports on the CPU to add four poles, so the CPU model would have to be changed. This would mean starting the design from scratch again, even though she had worked hard to minimize the cost increase by adding functions to the current model.

Manager Yoshioka spoke for Yuri, who swallowed the words.
"The specifications have already been approved, but they want to change them. However, the design has already been completed, so there is a scheduling problem. What about that?"
Katakura also had his position as a sales representative.
"The customer has its reasons, too. Once the specifications have been decided, we can't just say we won't change them. We can't change the schedule either. The client is sorry, but they are asking us to do something about it."
No kidding! You mean, "Change the specs, but don't delay the schedule!" Yuri shouted in her mind. But she didn't say it out loud. It was useless to

say it anyway.

"We're in trouble," Yoshioka said slowly.

"It's a problem for the control board, but what about the software?"

"It's a minor modification, so it's not a problem," Shimojo, who was in charge, answered to Yoshioka without hesitation.

"What about the outer packaging?"

Takeda seemed a little worried when asked but answered as if to himself, "It will probably only require a minor change to the mold, so I think we can manage the schedule."

"What about the radio? The radio module board won't be affected, but what about giteki?"

The person asked was Kirishima, the only project member who belonged to Development Section 1. All the other members belonged to Development Section 2.

"The inspection for giteki is just before shipment, so there is no problem. However, if the control board configuration changes significantly, we must change the application documents."

Giteki is a procedure that allows users in Japan to legally use this radio equipment without obtaining a radio station license, which in Japanese stands for Technical Regulations Conformity Certification.

"So only the control board is having the issue. Ueda-san, can't you do something about it? Can you at least consider it first?"

"In Japan, we say that the customer is God."

Shibata, the project leader, said to follow up. Then Iwata, the head of the Development Department, spoke up for the first time.

"The customer is not God. Think about it. When we make a wish, we put money in a money box at a shrine and ask God for a favor. In business, the customer pays, and we provide the product they want. Which is in the same position as God, the customer, or our company?"

Iwata, Director of the Development Department of the Equipment Division, was stocky like a barrel and had large, kind eyes. His playful tone brought a smile to everyone except Yuri's, and the atmosphere was relaxed.

"And God never complains."

His eyes turned sharp, and he now spoke in a severe tone. Everyone's smiles faded, and their faces returned to seriousness. Director Iwata again said in a gentle tone.

"Ueda-san, If you just can't do it, I don't blame you, but I think you can handle it. You have the technical skills and can come up with ideas."

"Uh, ……"

Yuri clammed up. She'd like to say she couldn't do it, but being praised by the director, Yuri didn't feel bad. Not only that. If she told him she couldn't do it, he might take it as denying her ability. She had to avoid that.

Under this company's performance-based system, the degree to which goals are achieved and the height of the goals were evaluated. Essentially, goals are set once every six months and do not change during the year, but the reality is different. The market, customers, technology, and trends of other companies change daily. Goals must also be changed flexibly. Refusing to respond to customer changes is equivalent to lowering targets, which will lead to a lower evaluation.

Director Iwata asked gently.

"Or is it impossible?"

"No, I don't mean impossible."

Yuri had to answer that.

"I know you can handle it. I'm counting on you. Besides, if you really can't,

I'll go to the customer and apologize."

Hearing these words, the words came out of Yuri's mouth reflexively.

"I will! I'll do it somehow!"

Oh shit, she thought, but it was too late.

"That's the spirit."

Director Iwata smiled at her. Yuri now felt like breaking out in a cold sweat. Yoshioka took over and said admonishingly.

"Ueda-san, we're asking you to do the hard issue. But the client is saying this because it is necessary. This is also where you can show your skills as an engineer. Please use all your wisdom. I'm also counting on you."

"Yes……." Yuri replied weakly.

Yoshioka said to all of them,

"What else is there to say? Good. Then, we will first consider adding a DIP switch ASAP. Depending on the results, we will consider how to proceed. Other than that, there are no other issues, is that correct?"

Director Iwata and the other attendees all nodded except for Yuri.

A picture of the product's white plastic case remained on the screen. Gazing at it, Sato, the Development Section 1 manager, murmured, "The new white model..."

Thus, the design review for Yuri's first product, in which she was in charge of the control board, ended. Just as Yuri returned to her desk, the chime for lunchtime rang.

Narumi Electronics Industry was a medium-sized electronic equipment manufacturer. The company had two divisions: the Equipment Division, which manufactured and sold communication-related equipment, and the System Division, which handled system integration.

The Development Section 1 of the Equipment Division's Development

Department was in charge of the wireless part of the business and had a small staff of four, including the section manager. Section 2, to which Yuri belonged, was a large section with 21 members in charge of non-radio parts, mainly digital circuits and software. Yuri was in charge of digital circuit design.

Although Yuri said she would do it somehow, she did not have a plan. Sipping her soba noodles in the company cafeteria, Yuri regretted her words.
"I'll do it somehow." was never the right thing to say up there. *"I'll do my best with it."* would have been fine …… Oh, no, I did it again.
Yuri hated her personality, tending to be stubborn. She had no appetite. She couldn't even hold her chopsticks with strength.
"Hey, you're unusually dressed in a suit. Are you going out today?"
Ayaka Nanase, a Yuri peer, sat in front of Yuri with a tray of A-lunch set on the table.
Ayaka was a software engineer in the Development Department of the Systems Division. Yuri and Ayaka were the only two peers in the company. Ayaka had an open personality, and because they were peers, she was the easiest person in the company for Yuri to talk to. They were now best friends, and Yuri did not have to worry about her words. Ayaka often made Yuri's tone more broken than usual as she accompanied Ayaka.

"No, no," Yuri answered weakly.
"Yuri is tall and slim, so it suits you. I envy you."
"You say that a lot. Men like a petite, glamorous girl like Ayaka. Besides, you look cute in that outfit."
Yuri gave her customary reply. Ayaka wore a company-issued blue work

shirt over a light pink blouse and a knee-length floral skirt underneath. It was cute but not showy and in good taste.

"Really?"

Ayaka seemed happy as she said.

"Yeah."

Yuri answered lamely and did not say the rest of the words.

Your eyes are flashing, and your naturally permed hair is nice and wavy—Ayaka's cute.

But the look on Ayaka's face said, "Say more."

Here it was. Ayaka's face was in maiden mode.

It can't be helped. I'll just say the last word.

"Ayaka is cute."

Ayaka heard that and said with satisfaction.

"Thanks. Yuri is pretty, too."

"I often get that a lot."

When Yuri joked back with no force, Ayaka's expression changed.

"What's wrong?"

"Nothin'."

Yuri changed the topic.

"I wore a suit today because a design review was held."

Ayaka nodded with a look of understanding. Her face had switched to business mode.

"Design review. It's the first product you designed. But it's an internal meeting, so you don't have to wear a suit."

"It's OK. I was betting on today."

Sadly, it was in the past tense, but she couldn't say that to Ayaka.

"You're fired up."

"Yes, I am. It's been almost two whole years since I joined the company.

I've been doing menial jobs and haven't had any accomplishments to show for it. It's like I'm being paid to study. We're being evaluated on a performance basis, so it won't be good for me if I don't produce results soon."

Yuri knew that although she had said so, her voice lacked power.

"Hmmm."

Ayaka's tone sounded like it was none of her business, even though she was in the same position.

Since Yuri said she would do it somehow, she had no choice but to consider it. This young engineer thought it over all afternoon. But no matter how she checked it, the number of input ports was insufficient.

A simple solution would be to add ICs to the control circuit to increase the number of input ports. But there is not enough board space for that. There is no space on the board for adding ICs in the pattern of components and wiring, at least around the DIP switches.

If she redoes the pattern design of the entire control board, there may be enough space. However, redesigning the pattern of the whole board takes time. Moreover, she might find it impossible after spending a lot of time studying the pattern. However, if the CPU were changed, there would probably be enough space.

Yuri had reached an impasse in her thinking. Should she change the CPU and redesign it from scratch? No, "I'll do it somehow" would be a lie. Changing the CPU would delay the schedule, which would not mean Yuri had done it.

When Yuri came to, it was almost seven o'clock at night. Usually, she would work a little more overtime, but it was unlikely that she would come up with any ideas if she stayed in the office to think about it. This young engineer's brain was getting tired. Yuri looked up at the ceiling and sighed.

Then, Manager Yoshioka approached Yuri.

"Ueda-san, why don't you go home today? You can think about it tomorrow. If you sit at your desk, you won't get any ideas."

Yoshioka was right, she thought. At this time, there was nothing to be gained by sitting at a desk. It was better to go home early and rest her head.

"Yes, I will."

Saying that, Yuri decided to leave.

"Good night."

"Good night."

Yuri left the office, thinking Manager Yoshioka worked harder than she did.

Yuri tried to organize her thoughts calmly on the bus and train home. It was confirmed that the number of input ports was insufficient. But the CPU had something in reserve: AD converters.

The CPU used in this project had four pins that could be used as AD converters. However, in the board that Yuri had designed, these terminals were used as digital signal inputs to input the values of the DIP switches. In other words, a single terminal was used to input a single-bit signal.

However, if it was used as an AD converter, eight-bit values could be obtained depending on the input voltage. In other words, 256 different values, ranging from zero to 255, could be obtained.

I would like to devise a way to read the status of multiple DIP switches with a single AD converter pin.

When Yuri arrived home with a faint hope, she looked at the clock on her mobile phone and saw it was eight o'clock.

Yuri's house was a single-family home in a suburban residential area where she lived with her parents. She could not find her father's commuting shoes when she entered the front door. It seemed that her father had not returned home yet.

"I'm home."

"Welcome home."

Yuri had not even finished her sentence when she heard her mother's voice coming from the living room.

Yuri went up the stairs and changed her clothes in her room on the second floor, then went downstairs to wash her face in the bathroom and went to the living room.

Yuri's mother was preparing dinner at the dining room table. There was grilled fish, boiled spinach, salad, tsukudani (preservable food boiled down in soy sauce), rice, and miso soup with many vegetable ingredients. There was enough for two. Her mother hadn't had dinner yet either. She probably waited until eight o'clock.

"Welcome home. You are earlier than usual today. I'm glad. We can have dinner together today."

Seeing her mother's face as she told her so was a relief. Suddenly, Yuri felt hungry.

"Itadakimasu," Yuri said. It's the Japanese expression of gratitude before meals. She sipped the miso soup and put her chopsticks to the grilled fish.

Yuri had a sister, Mana, who was two years older than her but got married the previous year and left home. Her mother worked part-time during the daytime on weekdays but was alone when she returned home in the evening. Yuri and her father came home late, so they often had dinner separately on weekdays.

Yuri's father, like Yuri, was a company employee and an electronics engineer. Previously, her mother used to wait until Yuri or her father returned home before having dinner. However, as she gained weight, her father told her to eat before them, and she did as he said.

"I'm glad I get to have dinner with Mom on a weekday," Yuri said honestly.
"I'm glad, too. Now, if only Dad could go home early."
Yuri continued to eat without answering her mother's words. She didn't dislike her father. But ......
"Ah!"
Yuri exclaimed involuntarily.
"What's wrong?"
Yuri's mother looked at her with her bowl of rice and chopsticks.
"No, nothing."
The daughter hurriedly finished her meal.
"Gochiso-sama," Yuri thanked Mom for the dinner.
"So fast."
Her mother seemed disappointed, unsurprising since she had been so happy to have dinner with her daughter.
"Sorry."
Yuri put her dishes away in the kitchen and entered her father's room.

The room had bookshelves lined with many books. Most of them were technical books and magazines related to electronics, and there were also science fiction novels and business books. Most books were in Japanese, but quite a few were in English. Yuri took out a technical magazine, a "Special Issue on AD/DA Converters," from among them and returned to her room.

To read the state of several switches with a single AD converter, that is, to make a DA converter with several switches. If the states of multiple switches can be converted to different analog voltages, the analog voltages can be read by the CPU's AD converter. By looking at the DA converter circuit in this book, she might have been able to find a way to do it.

The book says the most common circuits used in DA converters are the weighted resistor method and the ladder resistor circuit.

I can use these circuits.

A light of hope lit up in her heart.

However, upon closer reading, she couldn't be pleased. In both circuits, the digital signal is supposed to be a high-voltage or low-voltage input signal. But what was needed now was a circuit that creates analog voltages from the switch's states, turning on and off.

If a buffer IC is placed next to the switch, the DA converter in the book can be built. However, this would require too many components, and there would not be enough space on the control board for additional circuits. It would also increase the cost.

The light of hope was extinguished without a trace. A simpler circuit would be required to create an analog signal directly from the switches. Yes, simply and directly.

"It's not going to be easy," Yuri sighed. At that moment, she noticed that she could hear her father's voice coming from the first floor. Yuri looked at the clock and saw that it was ten o'clock. It was earlier than usual, but she decided to bathe first. She had come home early today, after all.

Yuri went downstairs with a change of clothes and shouted toward the living room.

"Mom, is the bath ready?"

"It's ready."

The words came back as expected.

While soaking in the bathtub, Yuri was thinking.

How can different analog voltages be generated, simply and directly, from each switch state?

Reading the state of multiple switches with a single AD converter is a generalized expression of the problem. In this case, it is enough if two switches can be read by one AD converter. If possible, the current CPU can read four more poles of DIP switches. The states of two switches mean two bits of information or four states. If limited to that, she might be able to make it work.

When Yuri got out of the bath and dried her hair, it was almost 11 o'clock. She decided to go to bed early today. Yuri put the book she had borrowed from her father's room into her commuter bag, turned off the light, and went to bed.

It would be better to go to bed early and think about it with a clear head tomorrow. Even so, Yuri had an image of two switches and one AD converter input terminal in her head. By adding a few more components, she wondered if it would be possible to create four different voltages corresponding to the state of the switches. The other components to use besides the switch would be resistors. The image of two switches and a few resistors drifted through her mind like a spacewalk.

Wait. Even though there are four different voltages, not all four need to be on the way from zero volts to supply voltage. One can be at zero volts and the other at the supply voltage. Only the other two voltages need to be on the way. And for that, ...

When Yuri was about to fall asleep, the moment came. Yuri opened her eyes.

"I can do it."

The young engineer felt like she could see everything. She jumped out of bed, turned on the light in her room, and went to her desk. Yuri got a piece of paper and a pen and began drawing a schematic.

First, there is an AD converter input terminal on the right side. From there, a signal line runs to the left. From the signal line, one line each runs up and down, and the upper line connects to the power supply via a switch. The lower line also goes through a switch and is connected to ground or zero-volt potential.

If the upper switch turns on, the voltage of the analog signal is equal to the supply voltage. If the lower switch is turned on, the voltage will be zero volts.

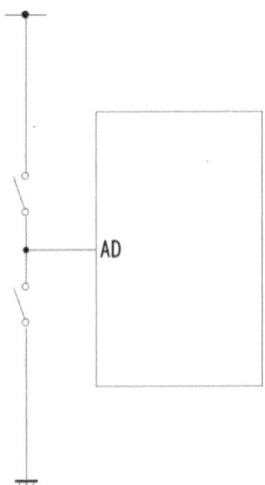

Of course, this is not the correct if as is. If both of the two switches are off, the potential is undefined. Besides, the power supply and ground will be shorted if both switches are on.

But Yuri already had the answer in sight. First, a one-kilo-ohm resistor is put in series with the upper switch. Then, a resistor is also put in series with the lower switch, and the resistance value here is two kilo-ohms.

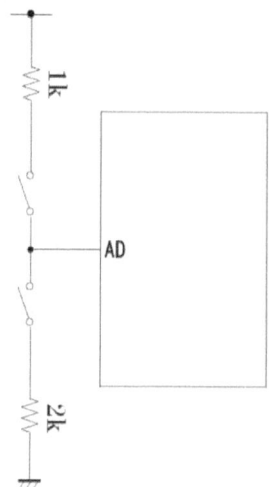

Thus, when both switches are turned on, the two resistors divide the voltage, and two-thirds of the supply voltage is added to the AD converter.

Additional resistors are needed to accommodate the case where both switches are off: a horizontal signal line from the AD converter input terminal is connected to the power supply through a two-megohm resistor on top and to ground through another one-megohm resistor on the bottom.

Now, when both switches are off, the supply voltage is divided by the upper two-megohm resistor and the lower one-megohm resistor, and one-third of the supply voltage is input to the AD converter.

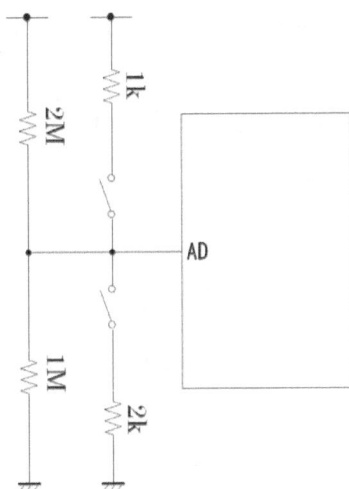

"In this way, when one or both switches are turned on, the one-kilo-ohm or two-kilo-ohm resistor is three orders of magnitude smaller than the one- or two-megaohm resistor, so its effect is dominant. Thus, as before the one- and two-megaohm resistors, if only the top switch is on, the voltage will be the supply voltage; if only the bottom switch is on, the voltage will be zero volts; and if both switches are on, the voltage will be two-thirds the supply voltage. Now we can give the AD converter four different voltages corresponding to the four different states of the two switches."

The following morning, Yuri explained the schematics she had drawn on the whiteboard to Manager Yoshioka, Shibata, and Shimojo. Today, Yuri is dressed as usual in blue work clothes over a simple sky-blue shirt and chinos underneath.

"I see. So the key to making four different voltages is to divide the power supply voltage into three parts instead of four," Manager Yoshioka was impressed.

"After reading the AD converter, the thresholds for determining the state of the switch are one-quarter of the supply voltage, one-half, and three-quarters. Is that correct? ...... No, it isn't," Shimojo has already started thinking about what must be done on the software side.

"The best thresholds are one-sixth, one-half, and five-sixths of the supply voltage. That maximizes the noise margin," Yuri replied.

As if reminded, Manager Yoshioka said, "Well, I think we should consider including resistance error and temperature characteristics as well."

"I have already considered it. It's OK, including resistance error and temperature characteristics."

"Really? Did you bring back the resistor datasheet? But you didn't get the idea until you got home yesterday, right?"

"I didn't get the idea until I got home yesterday. I looked up the resistor datasheets online."

"Oh, I see. We have that now," Manager Yoshioka said sincerely.

Then, in a firm tone, he added.

"It's wonderful, Ueda-san. As expected."

With these words from Manager Yoshioka, Yuri felt an indescribable sense of accomplishment and delight.

Manager Yoshioka added two more points.

"But one or two megohms seems a little too big. I think a hundred kilo-ohms or even a few dozen kilo-ohms would work, so consider it. Also, I think this might be patentable. Check online, and if you can't find a prior application, submit a patent application request to the Intellectual Property Division."

"Yes, I will."

Yuri explained the circuit in a meeting area near the department director's and section manager's desks. There was a table, chairs for four, and a whiteboard. The whiteboard could be seen from the seats of Iwata, the department director, and Sato, the manager of Development Section 1. Both Director Iwata and Manager Sato were looking at it from their seats with expressions of admiration. Yuri thought inwardly that she had scored some points. Director Iwata and Manager Sato looked at each other and nodded silently.

Yuri first confirmed that the larger resistances' values seem OK even if set to several tens of kilo-ohms. The noise margin might be the problem, but experimentation can only confirm this.

Next, she searched the database on the Patent Office website but could not find any prior applications. Putting off writing a patent application request, she conducted a confirmation experiment first.

She cut the wiring pattern on the board and wired the resistors and DIP switches by aerial wiring. She also asked Shimojo to modify the software for checking.
"Ouch!"
Yuri exclaimed involuntarily as she felt intense heat in the fingertips of her left hand while soldering in the wiring process. She was burned. The young engineer was holding the leads of the resistors she was soldering with her bare hands. It was no wonder she got burned. She had to pick them up with tweezers. But Yuri was in a hurry and worked with her bare hands. This had happened before.
"Well, I did it again."
Yuri put down the soldering iron and went into the hallway to cool her burned fingers with cold water from the water cooler.

The confirmation experiment went well. However, Yuri found that a capacitor must be placed between the AD converter input and the ground. Otherwise, the input voltage would change due to the charging and discharging of the sample hold capacitor inside the CPU.

On that day, she changed the schematic and asked the contractor to change the pattern on the board. She also submitted a summary of the patent application request form.

She does not need to rush to order the necessary parts, such as DIP switches, resistors, and capacitors, because they are in stock at her workplace. The parts list can be changed tomorrow.

The only thing left was to wait until the board with the changed pattern was ready. As usual, Yuri worked overtime that day until after 8:00 p.m. and left the office.

Yuri licked her burned finger as she got off the train and walked home along the shopping street in front of the station. It still tingled a little.
"Well, I did it again."

When she arrived home a little after nine and finished dinner alone as usual, her father came home and said, "Are you doing good?"
"Doing best!" Yuri unexpectedly retorted in a harsh tone and went back to her room.

The truth behind the suspicion

A few days later, in the afternoon, Yuri was fighting sleepiness. Perhaps it was because she was relieved that the arrangements for the circuit change had been completed, but she couldn't help but feel sleepy.
Oh no, I'm going to fall asleep. I'll have a cup of coffee.
Yuri got up and headed for the rest area.
At this hour, there shouldn't be anyone in there. I'll get a quick coffee and return to my desk. If anyone spots me, one might think I'm skipping work.
But she heard talking as she approached the rest area.
A meeting over tea?
Yuri stopped when she heard the conversation at the corner before the rest area because the story was unexpected.
"Give her to me."
Although the person speaking was not visible, it was the voice of Manager Sato of Development Section 1.
"No. I will never give her to you."
This is the voice of Manager Yoshioka, Yuri's boss.
"I need her."
"I do need her."
Oh, my God. I can't believe they're fighting over a woman. They both have wives.
Yuri returned to her desk, trying not to be noticed. Her sleepiness had all but disappeared.

That evening, Manager Yoshioka summoned Yuri to the conference room. Yuri and Yoshioka were the only two people in the room. Did he know Yuri had heard about the conversation in the break area? No, that was not possible. Yuri was listening to the conversation outside of the rest area and had not told anyone about the conversation. There was no way she could have told anyone. Yuri pretended she hadn't heard anything, but her body was tense and stiff.

Manager Yoshioka spoke up.

"I'll talk straight to you," he said. "Ueda-san, I want you transferred to Development Section 1."

For Yuri, this was unexpected. Her body was at once relaxed. However, being transferred to section 1 means she will be in charge of the radio-frequency circuits. Once again, Yuri's body was tense.

"Why? Why do I have to get transferred to section 1?"

"Section 1 is understaffed."

"But we're understaffed in Section 2, too."

Yuri knew that even though she said so, it was useless to say that.

"I don't want to give you up either. But I need someone to go. Or perhaps you can't develop radio-frequency circuits?"

"I can."

The usual strong words came out of Yuri's mouth.

"You can do it, right?"

"Yeah, well, ..."

Yuri just said that but could not continue.

The truth is, I can't.

Yuri thought, but Manager Yoshioka said, "I see. Well, I'll have to let you get transferred, as I thought."

"What do you mean, as you thought?"

"Well, it seemed that no one else could do radio-frequency circuit design, and Manager Sato and Director Iwata wanted me to do it. Sato-san told me to do it, and I turned him down many times, but Iwata-san made the decision himself."

Yuri had a flash of realization. That story she had heard in the break area had been about her. She hated herself for having misunderstood the situation so strangely, and her strength drained from her body.

"See, they are impressed with the idea of reading two DIP switches with one AD converter and thought you are good. It would have helped me if you had said, 'I can't design radio-frequency circuits.' But oh well, I guess we'll have to do it."

Oh my God, she thought.

Yuri could not say anything, feeling that she could not design a radio-frequency circuit, but she also felt ashamed of herself for her trivial misunderstanding.

On the train ride home, Yuri felt weak and regretful.

Why did I say, 'I can'? I want to design digital circuits, and I can't design radio-frequency circuits.

As she got off the train and walked home from the station, Yuri mumbled to herself.

"Oh, I did it again."

The next day, Yuri had lunch with Ayaka in the company cafeteria. She told Ayaka about her transfer to Development Section 1, but she hid from Ayaka that she was not confident in her radio-frequency circuit skills.

"Oh, you're moving to Section 1. ...... By the way, Kirishima-san in Section 1 is pretty cool, isn't he?"

For Ayaka, it didn't matter. So she was carefree.

"I wonder if he's cool."

"He's cool. What's he like?"

"I don't know. I've barely talked to him."

"Hmm."

After saying that much, Ayaka's face switched to business mode.

"I heard that Manager Yoshioka is going on another business trip to an overseas factory tomorrow."

"Really? I don't know anything about that. How do you know?"

"I've got a network of contacts."

Ayaka smiled wickedly, but Yuri felt sorry for Manager Yoshioka.

"Another overseas business trip? He's been on a business trip for months and returned the other day."

"I heard it is the final meeting for the issue."

"Which issue?"

"Don't you know? It's a quality problem. The customer was unaware of it, and there was talk of keeping it quiet, but the president decided to report it to the customer and take immediate action. They've come to a technical conclusion, and the rest seems to be a business problem."

"But why Yoshioka-san? That quality problem is caused by the production line, right? Is it a design problem?"

Yuri had heard that yields at overseas factories were worsening, but she had not heard anything about the quality of shipped products or design problems.

"It was a problem on the production line, but they couldn't solve it alone on the production floor. Even I don't know all the details."

Ayaka said regretfully.

"Well, but it's not so bad that our company is honest with the unaware customer."

"Yuri, you are so serious."

"It's not about that," Yuri said. "Besides, the risk of hiding a problem is greater nowadays."

"I see. You are right."

Ayaka nodded in agreement.

In the afternoon, Yuri received a file from the intellectual property section regarding an application request.

"That's weird. I've never received this file before the verbal description," she thought.

When she opened the file, it was disappointing. The first sheet of paper read, "Due to publicly known technology, we have determined that the application should be held in abeyance. The prior art application is attached."

After the second sheet, there were documents for the prior patent, and behind those documents was the bound application request form written by Yuri.

According to the prior application documents, the prior application was registered as a patent. In addition, a public notice bulletin was attached. The bulletin showed that the contents were the same as Yuri's idea.

"This is no good. We can't implement this." Yuri said to herself.

Yuri was impatient. She wanted to talk to Manager Yoshioka, but he seemed busy, so she consulted Shibata, the project leader.

"Is the content the same?" Shibata asked Yuri.

"Yes."

"And if we implement it, we infringe on it?"

"Yes."

"I see. ......, that's a problem."

Shibata looked at the public notice bulletin of the prior application with a serious expression for a while but suddenly became cheerful and said.

"The patent expires on July 5 of this year, and the delivery is on July 10, so it's all right."

"What?"

When Yuri looked closely at the documents, she found the application was twenty years old. Because it was an old application, Yuri's database search did not catch it. Lucky for them, they could implement it.

"Even so, I'm surprised that someone thought the same thing twenty years ago," Yuri thought.

"I'm glad we can implement it, but I'm sorry you can't get the patent." Shibata was right.

That night, Yuri was in her room at her computer. She had ordered several books on radio frequency from an online bookstore.

"At the very least, I must do some prep work," she said to herself.

The following Saturday, Yuri was in her room, reading and studying books on radio frequency. She didn't have much time, and there was no way she could understand everything. Still, Yuri wanted to understand at least a little, if not everything. Besides, she wanted to be able to say, "I have seen it before," for now.

What Yuri was reading was an explanation of decibels and matching. She could understand this. Next is the explanation of the Smith chart. Yuri had seen this "Smith chart," but she did not understand it. Yuri was puzzled.

The next day, Sunday, Yuri decided to take the plunge and ask her father to teach her. She felt uncomfortable asking her father, but it was better than asking someone at work.

Yuri's father asked her, "Before the Smith chart, do you know what a decibel is?"

"I do."

"Then explain it to me."

"Ten times the logarithm of the power ratio, or twenty times the logarithm of the voltage ratio."

"That's how you calculate it. What is the definition?"

"Definition? I just said that."

"You don't understand. OK, I'll explain it to you. It's essentially the logarithm of the power ratio."

"What about multiplying by ten?"

"The deci in decibel is a prefix, like deci in deciliter. Without the prefix, it is not decibel but Bell. Using the unit of the Bell, you don't have to multiply by ten. Just take the logarithm."

"I see. The logarithm of the ratio of power is Bell, and since it's the decibel, we multiply it by ten."

"Yes," he said. "So, do you see the reason why we use logarithms?"

"What? Reason?"

Yuri had never thought about the reason. She had only thought that such a unit of measure was actually used. As if he had seen through Yuri's thoughts, her father said to her,

"If all you're thinking is, 'That's just the way it is,' then you've got a bit of a problem. You're going to have to figure it out for yourself. Look, we use logarithms because the range of original values is so wide."

"Because the range of values is so wide?"

"Yes," he said. "For example, do you know what the range of power of a signal at the antenna of a wireless device could be?"

"Well, in the case of our company's products, the transmitter's antenna power is ten milliwatts at most."

"How would you describe that in dBm?"

"Uh, dBm is a decibel value based on one milliwatt, so it's 10 dBm."

"That's right. OK, what's the minimum power you'll use to receive?"

"Eh? To receive?"

Yuri could not answer.

"Well, that's OK. With the radio equipment handled by Yuri's company, the receiver's sensitivity would be about -110 dBm. What would be the range up to 10 dBm of transmitting output from this value?"

"Calculating the difference, 120 dB."

"120 dB means how many times the ratio at the original value?"

"Uh, 10 to the twelfth power, so ...... uh, ... times a trillion?"

"That's right. But with such a wide range, it would be inconvenient to use the value of 'one' or 'one trillion' as it is. It would be easier to say how many zeroes follow the number one, wouldn't it? And you calculate that by logarithms, right?"

"Oh, yeah! It's true!"

"But if you just take the logarithm, two times becomes 0.3. One-half becomes minus 0.3."

"Yeah."

"Perhaps the ratio of two times or one-half is used so often that they multiplied it by ten so that this would not be less than one, and that's how they came to use decibels."

"I see. Well, twice is plus three decibels, and one-half is minus three decibels."

"That's what I mean."

"I see. Then why do you multiply the logarithm of the voltage ratio by 20 when calculating decibels?"

"As I said before, it's essentially power, not voltage. Power is voltage squared divided by resistance, right?"

"Yeah, I know. The equation is 'P equals V squared divided by R,' right?" As Yuri's father listened to her answer, he took a piece of paper and a pen and wrote the following.

$$P = \frac{V^2}{R}$$

He said, "Yes. What do you get if you express the power ratio in terms of this right-hand side?"

"Uh, ..."

Yuri borrowed a pen from her father and wrote the formula.

$$\frac{P_1}{P_2} = \frac{\left(\frac{V_1^2}{R}\right)}{\left(\frac{V_2^2}{R}\right)}$$

"This is how it happens .... Oh well, the resistance goes away."

Yuri continued the equation.

$$\frac{P_1}{P_2} = \frac{\left(\frac{V_1^2}{R}\right)}{\left(\frac{V_2^2}{R}\right)}$$

$$= \frac{V_1^2}{V_2^2}$$

"What do you get when you take the logarithm of that?" Father asked. Yuri began to write the equation again.

$$\log\left(\frac{V_1^2}{V_2^2}\right)$$

"Got it!"
Yuri moved the pen further.

$$\log\left(\frac{V_1^2}{V_2^2}\right) = \log\left(\frac{V_1}{V_2}\right)^2$$

$$= 2\log\left(\frac{V_1}{V_2}\right)$$

"I see what you're saying. So when you calculate decibels from voltage, multiply it by 20, not 10."
"That's right. But you did delete the 'R' just now, right?"
"Yes."
"Why did you do it? Why could you do that?"
"Huh? Because the two R's are equal."
"Why can you say they're equal?"
"What? Not always equal?"
"Right. If the two R's are equal, you can delete them; if they are not, you can't."
"But are there actually cases where they aren't equal?"

"For example, when you express the amplification factor of an amplifier in decibels, the amplifier's input impedance is not necessarily equal to the output impedance, right?"

"Oh, I see."

"Although, they sometimes ignore the difference in impedance and use the voltage amplification factor in decibels. Well, then you'll have to learn the rest at the company. Don't just take what's in the books. Figure it out for yourself."

"Yes!"

Yuri turned to return to her room, but as she was leaving the living room, she stopped, turned around, and said.

"Thank you, Dad."

Dad seemed slightly surprised but soon said with an embarrassed smile.

"Hang in there."

"Yes, I will."

She wanted to say more than "Thank you, Dad." She wanted to say, "Dad, you're so cool," or "Dad, I love you," but she couldn't.

It had been a long time since she had felt this way. Yuri felt a confused and itchy sensation in herself for wanting to say such a thing.

Still, Yuri was happy. She was glad to understand decibels, but more than that, she was delighted to talk with her father, engineer to engineer. But then she remembered.

"Oh, Dad didn't teach me the Smith chart."

Transfer

On April 1, Yuri was transferred to Development Section 1. At the section's morning meeting, Manager Sato introduced Yuri to the section staff.

"Good morning. As you all know, Yuri Ueda-san has transferred to Development Section 1 as of today. We are a small section, and since Ueda-san is an excellent person, it is a significant increase in our power. I would like to welcome her."

Other than Manager Sato and Yuri, the three section members nodded.

"Today is April 1, but this transfer is not a lie. Just in case," Manager Sato added, and everyone smiled.

"Now, Ueda-san, a brief word of greeting, please."

Yuri, standing diagonally behind Sato, stepped forward to greet them.

"I'm Yuri Ueda. But, of course, I'm sure you know that."

Everyone smiled. Development Section 1 and Section 2 were in the same large room, so they had seen each other daily.

"I have never done radio-frequency circuit design before, but I will do my best to learn. I look forward to working with you."

Yuri thought she was like a new employee as she greeted them. The cherry blossoms were already in full bloom this year. In Japan, especially around Tokyo, cherry blossoms symbolize entrance into school and employment. However, there were no recruits in either Development Section 1 or Development Section 2 this year, and this year's recruits, who would be assigned to other departments, were still at their initiation ceremonies now.

It was a lie to say that she had never done radio-frequency circuit design before. But no one here knew that. Of course, they would think it is true. They would take these words as humility, not a lack of confidence, if they felt it was true. The words were intended not to lower their opinion or raise her expectations. However, Suzuki, a veteran of Development Section 1, had some surprising words to say.

"Just knowing that radio frequency is hard is a remarkable thing."

The word made Yuri think: did she raise their expectations of her? She didn't say it would be difficult. Was it on her face?

"Oh, um, ..."

Yuri was speechless.

"Then, I will ask Ueda-san to study for now, and I would like Kirishima-san to instruct her."

"Yes."

Kirishima responded to Manager Sato's words.

"Well, that's all for now."

Manager Sato declared the meeting closed, and everyone returned to their desks. Yuri also returned to her desk, which she had moved from Section 2 the day before. Next to her was Kirishima's desk.

Yuri was surprised to find that Kirishima was Yuri's trainer. She had expected to see Suzuki, a veteran, or Sugiyama, a mid-career worker. Yuri thought Kirishima was not yet a full-fledged engineer. He had not yet created a new design. Even for the product for which there was a design review in March, he had only made a few design changes to an existing wireless module.

Yuri thought she had become a full-fledged member, at least in Section 2. After all, she had been entrusted with designing a new control board for a new product. And yet, even though the educator is a year senior, he is not a full-fledged Kirishima.

Kirishima handed Yuri five books related to radio frequency and said, "First, study the books and ask me what you don't understand."
Two of them were the same as those she had bought, and she had already read them. She thought she was a little lucky.
"Yes."
Yuri answered, but she was still trying to be strong. In any case, she wanted him to stay with her to teach her.
  But she could not speak of such a thing. Especially to Kirishima. Yuri had no choice but to study on her own despite her anxiety.
"You better read this one and this one at the beginning."
One of the two books Kirishima mentioned was one she had already read.
"Yes."
Yuri answered honestly. There was no other way to answer.

  Yuri began reading the first of the two books Kirishima told her to read first, the one she was seeing for the first time. It seemed to contain only what she had already read in other books, so it was familiar to her. However, while some parts were recognizable, many more did not ring a bell. It was hard to imagine what the future held.

  When Yuri was eating with Ayaka during lunch break, she saw a group of black suits in the corner of the company cafeteria. Most other people wore blue work clothes, and the suits were only sparsely seen, so the group stood out.
  Ayaka saw this and said, "That's this year's newcomers. I'm jealous of those guys who got jobs in a seller's market."
"We had a tough time finding a job," Yuri replied.
"Last year, I went to the college I graduated and asked students to join our

company, then the professor said, 'With your grades if you had been job hunting this year, you could have gotten into any company.'"

Ayaka pointed at Yuri with her chopsticks and said, "You must be the same as me."

"Well, yeah."

Ayaka was right, but it didn't matter to Yuri now. Still, she added a word.

"But I heard the seniors had a much harder time."

"That's right. When my seniors were job hunting, there was a recession called the 'job-hunting ice age.' Compared to that, we were still better."

That evening, Yuri was invited to a welcome party at an izakaya (Japanese-style pub) in front of the station closest to the company. Yuri did not usually like to attend banquets at her workplace, but as expected, she could not resist attending the welcome party for herself.

"Ueda-san, why did you decide to become an electronics engineer?"

Manager Sato asked Yuri as he poured her a beer.

"Well, my father is also an electronics engineer, and that's how I got ……."

Yuri answered as vaguely as possible.

"I see. Does your father also develop electronic circuits for the company?"

"Yes."

"But he was tinkering with the circuits at home, too? If he is an office worker, you don't see your dad working, do you?"

He made an astute point. Yuri had no choice but to be a little more specific.

"My direct impetus came from finding my father's 'Denshi Block' when I was a kid, and I was intrigued by it."

'Denshi Block,' 'Electronic Block' in English, is a set of small plastic blocks with electronic components in each block that can be easily combined to experiment with various electronic circuits.

"Oh! Denshi Block. I miss it."

Suzuki leaned forward. Manager Sato reacted to this.

"Oh, Suzuki-san, did you use Denshi Block too?"

"Yeah, Sato-san, you too?"

"No, I was using 'MyKit.'"

"Oh, 'MyKit'."

Like 'Denshi Block,' 'MyKit' is a kit that makes it easy to experiment with electronic circuits. However, in 'MyKit,' the electronic components are fixed to the case, and the user is supposed to connect the components with wires.

Sato and Suzuki were grinning and pointing at each other, while Kirishima was puzzled. Yuri was relieved that the topic had left her. Sugiyama looked disinterested.

"Denshi Block and MyKit, I've heard of them but don't know much about them. In my day, it was computers."

Suzuki added a note to this.

"Don't misunderstand him, Kirishima-san. Sugiyama-san is talking about 8-bit CPU BASIC computers, not Windows or Mac computers."

"Yeah ......"

Kirishima simply answered that, but Sugiyama countered.

"No, Suzuki-san, It is PC-9801, not an 8-bit CPU! It's a 16-bit CPU! And Windows was released when I was in college."

"OK, OK. But it's '3.1,' right? Not '95.'"

Suzuki counterattacked.

"Yeah, well."

Sugiyama's advantage looked bad.

"Windows 98 was my first."

Kirishima finally joined the conversation.

"I started with an iMac. My grandma bought it for me when I was in high school."

Yuri was no slouch when it came to digital topics.

But then again, they are all techies at heart. But what about the radio?

"Sugiyama-san and Kirishima-san, how long have you been doing radio?" she asked.

"I didn't start radio until after I joined the company."

"Me too."

Sugiyama and Kirishima answered.

"I see."

It was unexpected for Yuri. Aside from Kirishima, she had assumed that Sugiyama had been doing radio as a hobby since he was a student.

"The days of the radio boy are long gone, right?"

Suzuki said to Manager Sato. Sato nodded.

They were a bit kawaii as they reminisced about their youthful days.

However, if both Sugiyama and Kirishima started radio after joining the company, it would be natural for Yuri to be able to do so from now on.

I wonder if I can make a radio-frequency circuit. She was once again starting to feel uneasy.

After a while, everyone was getting drunk. Suzuki was teasing Yuri.

"Radio-freeequency circuits aaare challenging, u know?"

"Ueda-san is an excellent engineer, so she'll be fine."

Kirishima, still calm, told him. Hearing this, Manager Sato said with a flushed face.

"Hey, Kirishima-san, Let's see if Ueda-san lives up to her reputation."

The phrase sounded familiar to Yuri. *Perhaps this is a GUNDAM phrase... or a Char phrase.*

But I will pretend I don't know. It will be troublesome later if I get into a conversation here.

As Yuri remained silent, Kirishima adjusted the conversation to Manager Sato.

"You aren't trying to bully your juniors, are you?"

"I nurse my share of grudges."

"Hey, I don't understand what you mean."

"Haha, yeah. ...... Well, let's call it a night."

Yuri's welcome party ended with a sanbonjime, hand-clapping for closing, with Manager Sato leading the group.

Smith chart

The next day, Yuri was concerned that Manager Yoshioka was taking a day off. She had heard that he had returned from an overseas business trip yesterday and would be coming to work today. They said he was not feeling well. Everyone said, "It's overwork." Yuri thought so, too. Manager Yoshioka needed some rest.

Yuri read the books given to her yesterday. She finished one of the two books she was told to read first and had already read the other book, which she bought herself. Yuri reread it, but it still didn't ring a bell. The most confusing part was the Smith chart.

The Smith chart is a diagram used for impedance matching in radio-frequency circuits.

In direct current, impedance corresponds to electrical resistance. In other words, it is the ratio of voltage to current. Since alternating current has amplitude and phase, the two quantities are expressed as complex numbers for impedance.

In radio-frequency circuits, the input and output impedances of two circuits must match when connecting them. Without matching, the signal power cannot be effectively transferred.

The Smith chart is a diagram in which several circles of different sizes within a large circle touch a single point on the far right.

A value of an impedance corresponds to a point on the Smith chart. If you put an inductor in series with that impedance, the point will move clockwise along the circle passing through that point. The distance to be traveled is determined by the inductance of the inductor. The point at which it moves corresponds to the impedance after the inductor is put in series. If a capacitor is put in series, it moves counterclockwise.

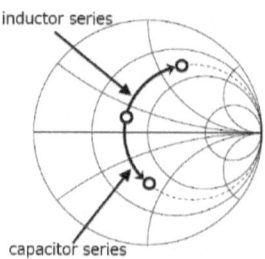

If an inductor or a capacitor is to be placed in parallel instead of in series, a diagram with several circles touching a single point at the left end should be used. In other words, the diagram looks as if the left and right sides are reversed. If an inductor is placed in parallel, the point moves counterclockwise; if a capacitor is placed in parallel, the point moves clockwise.

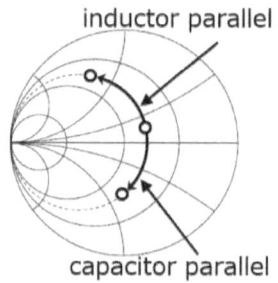

If she just used it as written in the book, there was no problem. Yuri could use it even if she did not understand it. But the young engineer wanted to feel that she understood. Besides, her father had told her not just to take what was in the books but to figure it out for herself.

Yuri read the section explaining the Smith Chart several times. She understood the written words and how to use the Smith Chart, but it didn't ring a bell. That was the end of her day.

The next day, Yuri arrived at work with one resolution in mind. She would ask a question today.

Kirishima had told her to ask about anything she didn't understand, and not only had Yuri finished the first two books she was told to read, but she had already read several other books on her own. Yet Yuri still didn't understand. No, it was not that she didn't understand. It was that the books didn't say what she wanted to know.

So, at worst, they wouldn't say, "Can't you understand such a simple thing?" Besides, she told them that she had never designed radio-frequency circuits before.

Yuri asked Kirishima at the right moment.
"Can you tell me about the Smith Chart? I know how to use it, but why does it do this?"
"What? What do you mean by 'why'?"
"I mean, why is the Smith Chart the way it is?"
"I don't know why you say that, ... but that's how it is."
"I can't understand 'that's how it is.' ......"
Yuri tried to resist modestly. Then Suzuki, who was listening, said.
"That's a good question. Kirishima-san, can't you answer it?"
"Well, even if you say why, what should I answer ......?"
Kirishima could not answer. Suzuki looked at him strangely.

"Huh? Didn't you understand?"

"What do you mean that I didn't understand? I understand how to use a Smith chart."

"No, what Ueda-san is asking is what the Smith Chart is in the first place."

Yuri, with powerful reinforcements, nodded broadly.

"In the first place?"

Kirishima said this and then fell silent.

"Uh, OK. I'll explain it to you, both of you. Listen, Smith chart is a unit circle of reflection coefficients with an impedance scale."

Yuri ruminated on Suzuki's words.

"A unit circle of reflection coefficients, with an impedance scale."

She thought she could see something a little bit.

Suzuki proceeded, "The absolute value of the reflection coefficient is between zero and one, and the phase angle is between zero and 360 degrees. So, if we represent it in a diagram, it would be a circle with a radius of one, that is, a unit circle."

Yuri nodded, ruminating further on Suzuki's words in her head and visualizing the unit circle. Suzuki went on to explain further.

"When you take the matching and make it 50 ohms, you take it to the middle point on the Smith chart. This means that the reflection coefficient is zero because the impedance at the center of the Smith chart, the reference impedance, is 50 ohms."

"I see!" Yuri exclaimed involuntarily.

"So, what matters when matching is the reflection coefficient, so you make a figure of it and assign an impedance scale to it. Then you can visually see the relationship between impedance and reflection coefficient at a glance."

"You're right. Excellent people ask different questions."

Suzuki answered with satisfaction. Kirishima was just impressed.

Yuri was filled with happiness, and she finally understood. But at the same time, she was surprised. Yuri had only asked a question that would not embarrass her, but to be called excellent was a lucky hit.

She asked one more question.

"But why is the reference impedance 50 ohms?"

This, too, was not mentioned in any of the books she read, so there should be no shame in asking.

"Uh, I can't remember," Suzuki could not answer.

Yuri suddenly noticed that Director Iwata had come to Manager Sato's desk and whispered something. Sato's expression suddenly turned grim. She wondered what was wrong. Sato nodded, and Iwata left the room with his bag. He seemed to be going out. It was already late afternoon.

That evening, Yuri asked her father a question. After dinner, she stayed in the living room, waiting for her father to return.

"Dad, why is the reference impedance fifty ohms?"

"Have one of the seniors in the office teach you."

Dad replied over a late dinner.

"I did, but he doesn't know."

"OK. Alright."

Dad put down his chopsticks and began to explain.

"The reference impedance is fifty ohms from the coaxial cable. A coaxial cable has a conductor wire, an insulator around it, a cylindrical conductor around it, and more insulation on the outside, right?"

"Yeah."

"The ratio of the inner diameter of the outer conductor to the outer diameter of the inner conductor determines the loss of high-frequency signals. If we draw a graph with that ratio on the horizontal axis and the loss on the vertical axis, we get something like this."

He said, drawing a flat "U" in the air with his index finger.

"Oh."

"Naturally, when determining the standard for coaxial cables, the ratio that minimizes loss was chosen."

"Well, that made the impedance fifty ohms."

"Don't be so quick to judge. That ratio alone does not determine characteristic impedance. It also depends on the dielectric constant of the insulator between the two conductors. Losses also vary depending on the insulator. Using polyethylene, a low-loss insulator, gives you a characteristic impedance of about 50 ohms."

"I see."

Yuri was pleased to know the reason and respected her father. But her father's explanation was not over yet. He continued talking.

"Do you know there is also a seventy-five-ohm coaxial cable?"

"Seventy-five ohms? On coaxial cable?"

"Yes," he said. "Before they used polyethylene, they didn't have a low-loss insulator, so they used air. When the insulator is air, the characteristic impedance with minimum loss is 75 ohms. There are still 75-ohm coaxial cables today as a remnant of that. Well, I guess the 75-ohm coaxial cable is now made with polyethylene by changing the ratio to make it 75 ohms."

"I didn't know there are seventy-five-ohm coaxial cables."

"There are even some measuring instruments that have an input impedance of 75 ohms. Well, most of them are probably fifty-ohms these days. Oh, but the TV antenna is 75 ohms."

"I see. Thank you."

Yuri hesitated momentarily but then said the words she wanted to say.

"Dad, you're amazing."

"Hm? Well, I've been doing this for decades."

Dad, who answered while holding chopsticks, looked large to her.

"I'll hang in, too."

"Yeah, but don't work too hard."

"Eh? You usually tell me to hang in there."

To Yuri's words, her father did not respond.

Oscillator circuit

The following day, when the first chime of the workday rang, Director Iwata said in a loud voice,

"Excuse me, everyone. Please gather in the center of the room."

Iwata stood up and raised his hands to call for attention. Yuri wondered what it could be about. Considering how he looked the day before, she didn't think it was good news. Everyone in Section 1 and Section 2 gathered around the director. Iwata said loudly and clearly so that everyone could hear.

"Manager Yoshioka was hospitalized. The cause is overwork. He will be absent for at least a week. It may be more than that."

Yuri's prediction came true. Instead, the news was worse than expected. She didn't expect him to be hospitalized. Is his condition that bad? A faint murmur spread through everyone.

"As for the travel report, I have received it via email and will forward it to all parties involved, but the issue has been resolved."

Perhaps he was relieved because the problem was resolved, and his previous exhaustion became a symptom.

"Regarding Manager Yoshioka's duties, I will concurrently serve as Manager, but Shibata-san will also perform some of them. I will manage attendance, and Shibata-san will handle practical matters."

Shibata nodded.

"Please avoid visiting him, as it will interfere with his rest. That's all for now."

Everyone returned to their seats. Director Iwata called Manager Sato and Shibata and went into the conference room.

Yuri picked up the third book, an oscillator circuits book. She was going to tackle this one next.

A short time later, Manager Sato returned to his desk. Although he appeared to be the same as usual, his expression was severe. Concerned, Yuri turned back and looked for Shibata in Section 2. Shibata had also returned to his desk, but his complexion was pale.

Yuri continued reading the book as if she had not noticed—no, she pretended to be reading. In the corner of her sight, Sato appeared to be looking at a computer screen, but his hands remained still. Manager Yoshioka's hospitalization would probably last more than a week. Yuri worried about Manager Yoshioka, but there was no use in worrying about him. The others had started their work. She, too, had regained her composure and began to read the book.

Yuri finished the book on oscillator circuits in three days. Rather, she spent three days reading it.

After all, the book's content was similar to what she had already read. At least she understood the logic. But ....

"Do you understand the principle of oscillator circuits?"

Kirishima spoke to her. It was as if he had been watching for the right moment.

"Yes, in a way. Amplitude condition and phase condition, right?"

Yuri replied, pretending to be modest.

"Yes. If the input signal of the amplifier circuit and the returned signal are in phase and the amplification factor is greater than or equal to the loss, it will oscillate at that frequency. It will then stabilize at an amplitude equal to the loss and amplification factor."

Yuri understood the meaning of Kirishima's words. It was precisely what was written in the book. However, it was still not entirely clear to her. As if Suzuki could see through Yuri's feelings, he stepped in.

"Your words are difficult to understand, Kirishima-san. Do you really understand what you're saying?"

"Eh? But that's what it says in the book as well. ......"

"Sure, the book says so, and it's true. What I am saying is, do you understand what it says? Try explaining it in your own words."

"In my own words?"

Kirishima was stumped. Yuri, too, was not sure what it meant.

"Yes," Suzuki said. "So that I can see what intuitive image you have."

"Intuitive image?"

Kirishima was still trying to understand what he was being asked to say, and the same was true for Yuri.

Is there more to it than what is written in the book? Since joining the company, Kirishima-san must have worked on radio-frequency circuits in Development Section 1 for three years. Since even Kirishima-san could not understand what was happening, Suzuki-san must have been asking about something quite advanced.

After all, Suzuki had been in this field for more than 20 years. For Kirishima, he was being humiliated in front of Yuri, the person he was teaching. Yuri felt sorry for Kirishima and became angry with Suzuki. Still, Yuri asked in an act of humility.

"Suzuki-san, could you please tell us what you mean?"

"Yes, I will. Well, let me explain it in my own words. Of course, you can understand it differently. That's all; I'm just saying that's how I see it."

Suzuki's calm tone made Yuri glad she did not speak harshly. Suzuki, keeping his tone calm, began to explain.

"First, let's narrow down the subject. There are various types of oscillator circuits. Here, we will focus on LC oscillator circuits, particularly Colpitts oscillator circuits. Crystal oscillator circuits are the same in principle."

Yuri and Kirishima nodded. Both of them had intended to do so from the beginning.

"Well, let's talk over there."

Suzuki pointed to the meeting table.

When Kirishima and Yuri arrived at the table, Suzuki drew the principal diagram of the Colpitts oscillator circuit on the whiteboard.

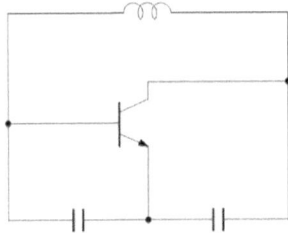

"This diagram shows up well in the book. Why does this circuit oscillate? In Kirishima-san's words, it is because the Phase and Amplitude conditions are satisfied. The logic is correct. But does that logic give you an image? It just doesn't ring a bell with me. So here's my understanding."

Yuri noticed that Sugiyama was also sitting beside her and listening to Suzuki.

"First of all, what is this transistor for? Of course, this is to amplify the signal. Without amplification, the signal will attenuate and eventually disappear. OK, let's assume first that there is no attenuation. If it does not attenuate, there is no need to amplify it. We don't need this transistor, either."

Suzuki erased the transistor.

"Now all that's left is one inductor and two capacitors."

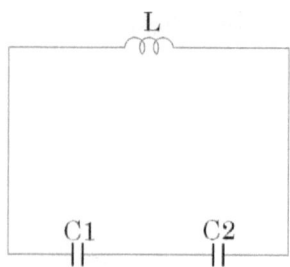

Suzuki named the inductor "L" and the two capacitors "C1" and "C2".

"Now, C1 and C2 are series connected. So this is equivalent to one capacitor C3."

Suzuki drew another diagram next to the first one, with one capacitor instead of two. He also added an equation.

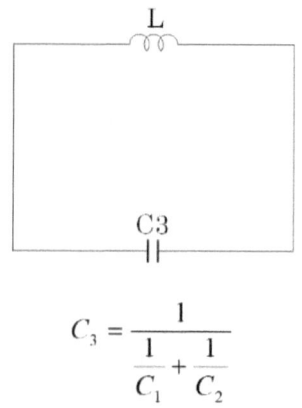

$$C_3 = \frac{1}{\frac{1}{C_1} + \frac{1}{C_2}}$$

"C3 capacity is obtained in this way, right?"

Yuri, Kirishima, and Sugiyama nodded.

"Now we have an LC resonant circuit with the transistor removed from the original diagram. What is the resonant frequency?"
Suzuki wrote without waiting for someone to answer.

$$f = \frac{1}{2\pi\sqrt{LC_3}}$$

"It is like this. The original oscillator circuit, which had the transistor, resonates at this frequency. Of course, we ignore the effects of the transistor here. If there is no attenuation in the circuit, it will oscillate at this frequency even without the transistor. Well, I should say that it continues to vibrate rather than oscillate."
Yuri, Kirishima, and Sugiyama listened in silence.
"Well, let me go back even further to the basics. I just said, 'vibrate.' How does this vibration happen?"
The three are stumped.
"Kirishima-san, what do you think? Why does it vibrate?"
Suzuki asked, and Kirishima answered in a rush.
"Uh, differential equations can describe the voltage and current in this circuit, and solving them will give us the vibrating solution."
"Correct. However, mathematically. I am discussing understanding it as an image, not a mathematical description. Then, let's think of mechanical vibrations. Then it's easier to understand because you can see it with your eyes."
Suzuki drew a picture of a spring. Its upper end was fixed to the ceiling, and the lower end was weighted.

"Pull this weight down and let go of it, and it will vibrate, right?"

The three nodded.

"As Kirishima-san said, this vibration can also be described by a differential equation, but how can we explain it in terms of a physical meaning rather than a mathematical description?"

Three of them thought about it.

"The hint is energy."

Hearing this, Sugiyama replied.

"Is it the spring's potential energy and the weight's kinetic energy?"

"Yes. When you first pull the weight down with your hand, your hand is moving while applying force. So, it is doing 'work.' In other words, it is giving energy to this system. By shifting the position of the weight, you are giving it positional energy ...... or, strictly speaking, elastic energy .... The potential energy here is the potential energy of the displacement from the neutral point of the spring. When the hand releases the weight, the weight begins to move with the force of the spring. In other words, it has velocity. So, part of the positional energy is converted into kinetic energy. Then, when the weight comes to the neutral point of the spring, the potential energy is zero, and all the initial potential energy is converted to kinetic energy. After that, the spring pushes the weight back, reducing the kinetic energy and increasing the potential energy. This is repeated to generate vibration. From the law of conservation of energy, the sum of potential energy and kinetic energy is always constant, and their ratio changes with time. This is the vibration of this system. OK?"

After confirming that the three nodded, Suzuki now drew a pendulum diagram.

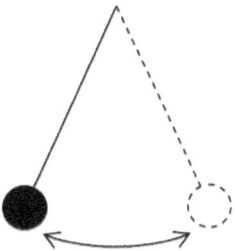

"Let me give you another mechanical example: a pendulum. The equation will be more complicated than the previous example, but you can see that the sum of kinetic and potential energy is constant, and the ratio of the two changes. Now, back to the circuit: the LC resonant circuit. This circuit vibrates in the same way. So, in this vibration, the ratio of what energy to what energy changes?"

"The capacitor's electrostatic energy and the inductor's magnetic energy." Again, Sugiyama answered.

"Yes, the essence of oscillation in an LC resonant circuit is a change in the ratio of electrostatic energy to magnetic energy. The form of energy is different, but the essence is the same as in the two examples of mechanical systems. Understand?"

Kirishima said impatiently, "Yes, but how does that relate to the oscillator circuit?"

"Well, just bear with me a little longer. We're almost there. To oscillate an oscillator circuit is to sustain the vibration of an LC resonant circuit. The vibration of the LC resonant circuit is essentially the same as that of a pendulum. So, oscillating an oscillator circuit is essentially the same as sustaining the vibration of a pendulum. OK?"

"Yes," Kirishima answered.

"Then how do you keep the pendulum vibrating?"

"Swing the hand holding the string?"

This time, Yuri answered.

"Great. But what else? Any other specific examples?"

The three of them twisted their heads.

"It's a swing. A swing on the playground. Swings are pendulums, too, right?"

"Oh, I see."

The three nodded.

"You've played on a swing, right?"

The three nodded again.

"That's the oscillator circuit."

"What?"

All three of them didn't know what he was talking about.

"Because on a swing, a person generates and sustains vibrations."

"Huh."

That being said, maybe so.

"Remember how you play a swing on a playground? How do you play a swing?"

"Like this ......"

Yuri pretended to grip the chains with both hands and swung both legs back and forth from the knees to the toes while sitting on the chair.

"Yes. You move your center of gravity back and forth by swinging your legs. This generates or amplifies vibration, which corresponds to the amplitude condition. But if you swing your legs blindly, it won't work, will it?"

Sugiyama hastened to add.

"Once the vibration starts, we swing our legs in time with it, in other words, in synchronization. So this corresponds to the phase condition."

"Oh."

Yuri and Kirishima were impressed. Suzuki nodded his head.

"That's right. So, an oscillator circuit is essentially the same as a swing."

Yuri, Kirishima, and Sugiyama nodded, this time broadly.

"Now let's consider how we can achieve the same thing as the swing with a circuit. But before that, suppose there is a swing with a person on it. What determines the period of that swing?"

"It's a pendulum, so it's determined by the length of the string, or in the case of a swing, the length of the chain. The mass of the weight, or in this case the weight of the person, or any other factor, is irrelevant."

Kirishima replied.

"Yes, it is. But strictly speaking, that's for small amplitudes. If the amplitude is large, the period also varies with the amplitude."

"Is that so?"

"Yes, it is. Well, actually, I only recently found out about it myself."

Suzuki smiled. The three of them were impressed.

"Well, let's assume that the amplitude is small enough. If so, the period of the swing is determined only by the length of the chain. But isn't there something odd about that?"

"Hm? But that's what a pendulum period is, isn't it?"

Kirishima answered. Yuri and Sugiyama thought the same. They didn't see anything odd about it.

"I just said that a pendulum and an LC resonant circuit are essentially the same, so what determines the resonant frequency of an LC resonant circuit?"

The three of them gasped. Yuri said.

"The two, L and C. But, strangely, the pendulum is determined only by the length of the thread."

Suzuki remained silent. Then Kirishima, who had been looking down and holding his head, looked up and said, "Oh, yeah. Gravity acceleration."

Yuri and Sugiyama were impressed and looked at Kirishima's face.

"Exactly," Suzuki said confidently.

"Then let's get back to how we can achieve the same thing with the swings but with circuits."

Yes, that was what they were talking about.

"A swing that a person does not swing is the same as a pendulum, and a pendulum is the same as an LC resonant circuit. If you 'swing' the LC resonant circuit by something, it will become an oscillator circuit. Swing it by what?"

"By an amplifier circuit."

Sugiyama replied.

"Right. If an amplifier circuit is made normally, it can have some frequency bandwidth. So, the amplitude condition is a small problem, except when the frequency is higher. The problem is the phase condition. It is important to note that the phase condition must be considered for the entire oscillator circuit. In other words, it is necessary to consider including an amplifier circuit. The simplest amplifier circuit can be made with a single transistor.

What is the phase characteristic of this?"

"Since it's an inverting amplifier circuit, the output phase is the opposite phase of the input, right?"

Sugiyama quickly replied.

"Yes, it is. On the other hand, the oscillator circuit, as a whole, needs to apply positive feedback, that is, feedback in phase."

The three nodded.

"So, what's the phase of the feedback circuit, excluding the amplifier circuit?"

"It needs to be in the opposite phase."

Yuri replied.

"Exactly. It goes into an opposite phase in the amplifier circuit, then further into the opposite phase, and feeds back. Then, feedback in the in-phase is applied. So, to make an oscillator circuit, an inverting amplifier with a transistor should be added to the LC resonant circuit so that feedback is applied in the opposite phase."

Suzuki once again drew a diagram of the principle of the Colpitts oscillator circuit.

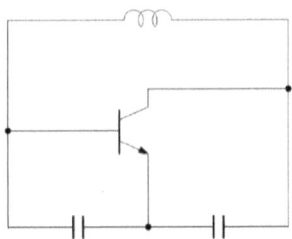

"As I said, these inductor and the two capacitors are an LC resonant circuit.

The two capacitors are there to divide the voltage. The voltage divided by the two capacitors is connected to the transistor emitter, right? "
The three nodded.
"If we look at the transistor amplifier as an emitter-grounded amplifier circuit, it operates with the emitter potential as the reference potential. So, let's assume the emitter voltage is zero."
Suzuki wrote "0" near the emitter.
"It's an inverting amplifier, so if the input, the base, is positive, the output, the collector, is negative."
He wrote "+" on the base and "-" on the collector sides.
"Then I'll draw arrows in the direction of the voltage, from positive to negative."
He drew an arrow from the base to the emitter and an arrow from the emitter to the collector.

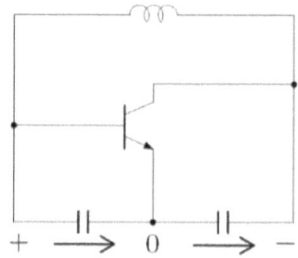

"Wow!"
Yuri shouted unintentionally.
"I see, looking at it from the LC resonant circuit, the arrows are in the same direction. That is, feedback in phase is being applied."

She felt her vision suddenly open up. Kirishima and Sugiyama nodded their heads.

"That's right. On the other hand, from the amplifier circuit's point of view, the feedback is in the opposite phase. Because the collector output is negative and the base input is positive."

Suzuki said, satisfied.

"So, when looking at this diagram, I think it is important to look at it from the two perspectives of the amplifier circuit and the resonant circuit and note that the amplifier circuit's reference potential is the voltage in the middle of the resonant circuit. This is my way of understanding the principle of the Colpitts oscillator circuit. Do you understand?"

"Yes!"

Yuri replied with a smile to Suzuki's confirmation.

"By the way, how did Kirishima-san see it?"

This time, Suzuki asked.

"Sorry?"

It was a question that neither Yuri nor Kirishima had expected. Suzuki repeated.

"I understand it as I explained, but Kirishima-san, do you understand it just by looking at the equation or the principle diagram? Or do you have a different way of understanding it?"

"Way of understanding……"

Kirishima could not answer. He seemed defensive, as if afraid he would be scolded. Suzuki asked further.

"You said you read the book and understood it. You mentioned amplitude conditions and phase conditions, and I'd like to know if you could give me an image of what you mean by that and what you take it to mean."

Yuri was annoyed that he did not have to be so sarcastic. But simultaneously, she remembered her father's words, "Don't believe everything you read." She was reminded that she had taken what was written in the book for granted. Until now, she had not understood what it meant to take things for granted. Yuri felt ashamed of herself after Suzuki confronted her with an understanding that was not taken for granted.

"I didn't take anything or image anything. I just took what the book said." Kirishima finally answered. Kirishima seemed to feel the same way as Yuri. Yuri nodded involuntarily and prepared for Suzuki to become angry.

However, Suzuki's words came as a surprise.

"Oh, really? I didn't know that."

Yuri, whose expectations were off, needed help understanding the word's meaning. Kirishima seemed to do the same.

"I thought you bright people would understand it easily just by reading a book because the people who joined the company during the ice age of employment are all excellent. Because people who would normally go to bigger companies came to our company."

Nowadays, the best and brightest may not necessarily go to the big companies, but Suzuki's question to Kirishima was not sarcastic but serious. Sugiyama nodded and said,

"That's right. But then again, even though I had read the book and understood the logic, it didn't ring a bell, but that's how you look at this diagram."

Yuri agreed with the words, "It didn't ring a bell." And she felt she understood a little bit more about what it means to really understand. It's not like writing the correct answer on a college exam. That's not enough.

"Um, Suzuki-san, did you come up with that explanation alone?" Kirishima asked fearfully.

"Yeah. There may be a book with the same information, but I haven't seen it. It would be easier to understand for me if it was written there."

"It's amazing."

"It's not amazing. But I kept thinking about it because it didn't ring a bell. I kept thinking, 'How should I understand it?'"

"Have you been thinking about it all the time?"

Kirishima was surprised. Yuri was the same.

"Well, I say 'all the time,' but I have work to do, so I can't think about this all the time. I thought about it from time to time while working between jobs. Recently, I finally figured it out. If I look at it this way, it rings a bell. Or should I say, it makes sense to me."

"So that's why I wasn't taught either."

Sugiyama was convinced.

"That's right."

Suzuki nodded with a smile.

    Yuri was relieved to learn that she was not the only one who had read a book that didn't ring a bell or make sense to her. She realized the importance of thinking and understanding for herself. It is no good to take it in faith. After all, veterans were different.

"Earlier, Ueda-san said 'in a way.' That means we still have a way to go. We need to understand more deeply. I hope you cherish that kind of feeling."

Suzuki seemed to misunderstand Yuri. She didn't have such an admirable mindset. She felt guilty.

Barrack circuit

A week later, Yuri finished the rest of the book. Director Iwata said nothing, but it seemed Manager Yoshioka was still in the hospital.

After all, many books she read did not ring a bell. But she was not alone in that. So, Yuri decided to try to understand without rushing. First, it would be good if she could understand the logic. Yuri felt that way.

"Well, can you build a circuit and get it working?"
At Kirishima's suggestion, Yuri decided to try making an oscillator circuit. Even though she would make a circuit, she would not design a pattern for a printed circuit board this time. She would use a universal circuit board, put the components on it by hand, and wire it. This is what is called barrack.

Of course, Yuri herself designed the circuit. However, she just decided on the connections of the transistor, crystal, and capacitor as described in the book, settled on the bias resistor, and then decided on those values to match the characteristics of the transistor. So, of course, it was supposed to work. But it was only "supposed" to work.

The making of the barrack circuit was finished. It took little time to make since it was a small-scale circuit. Still, she checked again and again. She checked that the design conformed to theory, that the values were appropriate, and that there were no mistakes in the wiring. Once that was done, Yuri connected the power supply and attached an oscilloscope probe to the transistor collector to observe the oscillation waveform. All that was left was to turn on the power.

Since moving to Development Section 1, Yuri had been dreading this moment. However, she knew that it would eventually come. Yuri felt like praying to God. She was afraid to turn it on. Her heart was pounding. The memory of that time came back to her mind.

When Yuri was in elementary school, she happened to find her father's old 'Denshi Block.' She became interested in it and joined the technology club in middle school, where she became hooked on digital circuits.

Yuri's advisor encouraged her to try making an analog circuit as well. Yuri followed the recommendation and built an oscillator circuit. It was not a digital multivibrator but an analog LC oscillator circuit.

The circuit was supposed to oscillate at about ten megahertz. But it did not work. After that, Yuri stopped working on analog circuits, especially radio-frequency circuits. "I can't do it," she concluded. As if to cover up her dislike of them, she became even more addicted to digital circuits.

Why didn't that circuit work at that time? She did not know that now. But this circuit needed to work. Yuri turned on the power. Then, the oscilloscope displayed a sine wave.

"Oh!"

Yuri's feelings came out in her voice unexpectedly. Yuri was relieved. No, surprised was a better word. She would be happy but not surprised if this were a digital circuit. But this was a radio-frequency circuit. This was the first time a radio-frequency circuit, which she had designed and built herself, had worked. Yuri was surprised to learn that radio-frequency circuits could also work if they were correctly made.

"It's nice when it works, isn't it?"

Kirishima, who was right next to Yuri before she knew it, said. He was looking at her. She was seen with a big smile on her face.

"Well, yes."

Yuri barely answered but felt like her face was about to catch fire.

It was 'Golden Week,' the May holidays in Japan. Looking out the window of Yuri's room, she saw that the azaleas in the sidewalk plantings were blooming beautifully.

Even though it is a consecutive holiday, it is difficult to go on a trip this year because of the 'series of holidays with some workdays in between'. So, Yuri decided to spend her time at home studying radio-frequency instead of going anywhere. Yuri would have done so this year even if she could have taken a long holiday.

Yuri went through more books she had bought. Suzuki misunderstood her. He thought she was better than she actually was. She felt she had to get close to that level.

After the holidays, when Yuri turned on the barrack circuit, it did not work for some reason.
"Eh, why?"
For the moment, she turned it off. She felt her face turn pale. *Why isn't it working?* It was working fine before the holidays. Yuri checked the wiring, and nothing was disconnected. The power connection was also correct. She turned it on again. It still did not work, so she immediately turned off the power. Yuri touched parts here and there, but no component was hot. She strained her eyes and looked at the board from top to bottom and left to right.
"Hmm?"
Yuri noticed a slight gap in the power cable connector. It was stuck partially in.
"Is this it?"

Yuri plugged the connector all the way in and turned it on again. Then, a spectrum appeared on the screen of the spectrum analyzer. Before the holidays, Kirishima had told her to use a spectrum analyzer instead of an oscilloscope.

Yuri was relieved that the barrack circuit worked adequately. She felt her pale face return to normal.

Yuri had added a frequency modulation function to this barrack circuit on the workdays during the holidays. Using it, she confirmed that changing the frequency of the modulating signal or the amount of frequency deviation would change the spectrum.

"Yeah, that's good. OK, now let's make a printed circuit board," Kirishima said after checking the oscillator circuit report written by Yuri.

Making a printed circuit board means designing a pattern for a printed circuit board and ordering a board fabrication company to manufacture the printed circuit board.

"Eh? I'm going to design a product?"

Yuri hurriedly asked back.

"No, no, no. The pattern characteristics are unstable with barracks, so you must make a PCB to learn impedance matching."

PCB stands for printed circuit board.

"Do we make PCBs just for the sake of learning?"

"Yes, we do. Here, install this on your computer."

Kirishima handed Yuri a CD-ROM with a CAD program.

"What about the license?"

Yuri asked to be sure. Without a license, it would be an illegal copy.

"Of course, we have one for you. I bought it when Ueda-san was transferred to Section 1."

"I got it."

"The next one you make will be a receiver. I will email you the data sheets of the ICs to be used."

"Yes."

A little later, PDFs of the data sheets arrived by email. They appeared to be old ICs. They were an IC with the circuits after the intermediate frequency (IF) stage on a single chip and an MMIC (monolithic microwave IC) for frequency conversion in the first stage.

"You can use them to build a receiver, and the circuit around the ICs should be the same as the recommended circuit in the datasheet. The local oscillator circuit should be the same as the one in the barrack. First, draw a schematic of the whole circuit, and then design the pattern."

The local oscillator circuit Kirishima mentioned is an oscillator circuit inside the receiver that is used to convert the received signal to a lower frequency.

As Kirishima told her to do, Yuri first drew a circuit diagram in CAD, and since it was the same as the circuit recommended by the IC manufacturer, there was no problem. However, she was told to add some inductors and capacitors in series and parallel for impedance matching in the high-frequency signal path.

Once the circuit diagram was completed, the next step was to design the pattern for the printed circuit board. Because it was a radio-frequency circuit, it differed from a digital circuit. With this in mind, Yuri drew the circuit board pattern, but Kirishima still gave her a few orders.

"The pattern for the high-frequency path was originally intended to be a fifty-ohm line, but that would be too wide. So please, let's make it 0.5 mm wide."

"Don't draw the pattern bent ninety degrees here. It's a high-frequency pattern."

"Bypass capacitors should have multiple capacitances, in parallel, and be connected to the ground by multiple through-holes."

"The ground can't just be filled in. You have to divide the ground into different areas, depending on which signal the ground is for."

"The pattern of a radio-frequency circuit board is not just a matter of following the schematic. You design it by imagining the wave traveling through the pattern and coming back through the ground."

Yuri had already read about many of these orders in books. She just got into the habit of doing them while working on digital circuits. Yuri had checked and fixed it quite a bit, but there were still some omissions. That was a little frustrating, but it was fresh and rather enjoyable.

When she finished her work, Yuri noticed several emails on her cell phone. All of them had the subject "Happy Birthday!" Well, it was Yuri's birthday, May 25. The emails were from her high school friends. The email with the subject "Rounded to 30, congratulations!" was from Ayaka.

Don't round up my age.

In addition, there was a congratulatory email from "Mana Ueda." Oops. Yuri had to change her elder sister's family name on her cell phone.

Finding the circuit values

In June, the prototype boards that had been ordered were ready. They were so-called "bare boards" before soldering the components. As Yuri was mounting—soldering—the components to the board, Suzuki brought a microscope. It was a stereo microscope, a type of microscope that you see with both eyes.

"You can use this."

"Oh, thank you."

Indeed, it was a little difficult to see with the naked eye because of the small parts used. Resistors, capacitors, and inductors were 1005 in metric, or 1.0 mm x 0.5 mm. Yuri continued soldering using the microscope.

Yuri picked up a 1005-size chip capacitor with tweezers in her left hand and placed it where she would solder it. Yuri moved the circuit board with her right hand so that the capacitor was in the center of the microscope's field of view while holding it with tweezers. Holding a soldering iron in her right hand, she put solder on the tip and looked into the microscope. In the microscope's view, a chip capacitor held in place with tweezers was visible in the center. To solder, she moved the tip of the soldering iron in her right hand to the capacitor. She intended to move it, but the iron tip was out of sight.

With a microscope, the area to be soldered appears large, but the area around it is obscured. The field of view becomes narrower.

"Huh? Where is the tip?"

At that moment, she felt intense heat on the index finger of her left hand.

"Ouch!"

Yuri exclaimed involuntarily. The tip of the soldering iron hit her finger. Yuri noticed Kirishima, Sugiyama, and Suzuki smiling and looking at her.

They must have had the same experience. Yuri hurried out of the room and cooled the finger with a water cooler in the hallway. Then she remembered—that she had been burned in middle school, too.

"Well, I did it again."

But again, when I was in middle school, ...

In middle school, Yuri taught herself to make digital circuits in the engineering club and was praised for it. But when she told her dad, he said, "Despite being a girl......," that's all.

She remembered a neighbor lady telling her that girls should cook, sew, or ...... like her sister. It was a shock to Yuri. Sure, her sister Mana was an excellent cook and seamstress, affectionate and girly.

But Yuri loved electronic circuits. She thought, why should she be told that "Despite being a girl" when she is doing her best at what she loves to do? So, she was determined and hammered into the digital circuit.

Then, her advisor told her also to try analog and radio-frequency circuits. But the oscillating circuit she built didn't work, and she hated it. So she decided to become as good as anyone else in digital circuits and not let anyone say, "Girls should..."

The components had been mounted and checked. Now, it was time to turn on the power. Yuri connected the stabilized power supply to the board.

"May it work!" Yuri prayed in her heart.

Feeling her heart pounding, she turned on the switch. The local oscillator circuit was oscillating correctly, which was a relief. Yuri sighed. The next step was to check if the receiving operation was regular. A signal generator generated a frequency-modulated signal on the receiving frequency, which was input to the receiver's connector for the antenna. Looking at the receiver's output with an oscilloscope, the signal was demodulated, although the waveform was somewhat distorted. For the time being, it worked. Yuri was relieved and impressed at the same time. She thought it would work if it was designed and built correctly.

Yuri was having lunch at the company cafeteria when Ayaka, next to her, said something strange.

"Hey, did you hear? Manager Yoshioka is rumored to have cancer."

"What? Don't be silly!" Yuri glared at Ayaka involuntarily.

"But that's what they say."

"It's just a rumor."

"But don't you think it's strange? Manager Yoshioka has been in the hospital for some time. We haven't been informed of his condition, and we're not even allowed to visit him."

Indeed, it is. Perhaps .... No, it couldn't be. Yuri drowned out the unpleasant thought. But Ayaka said.

"I wonder if maybe this rumor is a deliberate leak."

"What do you mean?"

"So, just to ensure everyone is prepared in case something happens."

Yuri was surprised to hear Ayaka think this way. She could understand the logic, but it would never happen. It couldn't be true. She felt the blood rush to her head.

"You overthink. Doubts beget doubts!"

Yuri stood up from her seat with her lunch tray. She ignored Ayaka's voice calling behind her. As she put the dishes in the return slot, she also returned the unpleasant thought. Besides, it would soon be the dreary rainy season.

The chime for the end of lunch break rang. Now, the real work was beginning. Yuri had made this board for practical training in impedance matching.

Matching is performed using a network analyzer. The network analyzer measures the reflection coefficient and displays it in graphs such as the Smith chart and return loss.

Yuri connected the network analyzer to the receiver's connector for the antenna and displayed the reflection coefficient in Smith chart format. The target was the center point, with an impedance of 50 ohms. However, the point displayed was far to the upper right of the center point. The standing wave ratio (SWR) was 13, the reflection coefficient was 0.86, and the return loss was 1.3 dB. In short, more than 70% of the power of the incoming signal from the antenna was reflected back.

However, the values of the inductors and capacitors for matching were to be determined. If Yuri could get matching with them, that is, if the input impedance got close to 50 ohms, it would be good.

At this stage, the matching elements had a 1000pF (pico farad) capacitor in a series connection. The first step was to reduce the capacitance of that capacitor, which would have made the impedance closer to 50 ohms. The computer calculates the capacity value. They use Smith Chart calculation software. According to the calculation result, about 3 pF seemed good.

Yuri replaced 1000pF with 3pF. Looking at the network analyzer, the point on the Smith chart had indeed moved significantly. But it was strange. It should have moved counterclockwise on the circle inscribed in the outer circle at the rightmost point. However, the direction of movement was different.

"Why?" Yuri thought to herself.

Yuri checked again to see if she had made any mistakes. But nothing was wrong. She was simply measuring the input impedance, so she thought there was nothing to make a mistake.

I'll just have to try something. OK, Let's try 10 pF.

When the capacitor was replaced with 10 pF, the point on the Smith chart moved to a different position again. If there was the 10pF point on an arc connecting the 1000pF point and the 3pF point, then that arc was not like the arc it should have been. When she measured with 6pF and 22pF, those points were still on that "different arc" than they should be. Yuri held her head in her hands. It didn't make sense. Her head was about to freeze up.

"What's wrong?"

Kirishima spoke to Yuri. She wanted to solve the problem independently, but now that she was spoken to, she had no choice. Yuri explained the situation by showing the points of the measurement results plotted on the Smith Chart calculation software. After thinking for a moment, Kirishima asked Yuri.

"Have you calibrated the reference plane?"

"Reference plane?"

"Looks like you haven't."

Kirishima explained the reference plane. First, the network analyzer measures the amplitude and phase of the traveling and reflected waves. Second, if the length of the connected cables changes, the reflected wave's phase shifts. So, we need to compensate for the phase to match the cable. Calibrating the reference plane means opening the end of the cable actually used and compensating the phase so that the measurement point comes to the right end on the Smith chart, the point that means open.

Yuri was depressed. Once she knew, it was obvious. Her incompetence was finally exposed. She had pretended to be brilliant, but it was all in vain. Yuri lost her energy. Seeing Yuri's state of mind, Kirishima said.

"Don't be discouraged. I've done it too."

"Oh? You too?"

"Yup, hey Suzuki-san, Ueda-san did it, too."

To Kirishima's voice, Suzuki replied in a joking tone.

"Well, now Ueda-san is a respectable half-fledged engineer."

"Respectable half-fledged?"

Yuri couldn't help but ask back. She had no idea what it meant.

"It takes ten years to become a full-fledged radio-frequency circuit engineer. Even a half-fledged engineer is impressive," Suzuki replied.

"It means the baptism," Kirishima explained.

Yuri still wasn't sure, but she was sure she wasn't considered incompetent. A little energy came back to her.

Yuri regained her composure, matched the reference plane, and resumed the matching process.

She restored the value for matching and measured it again. This time, a point appeared in the lower right corner. So, if an inductor was put in series, the impedance would approach 50 ohms. She left the capacitor in series with the MMIC's input terminal for frequency conversion as was. She replaced the provisional capacitor in series beyond it with a 10 nH (nano-Henry) inductor. The series capacitor was necessary because the MMIC's input terminal must be DC isolated. When measured, the point moved to a position almost exactly as predicted by the software. However, it is slightly offset.

Hmmm, well, I'll keep trying.

When measured with 22nH and 4.7nH, it was nicely on the arc. However, the arc was slightly offset from the intended arc.

"I wonder why. Earlier, the reference plane was not aligned. I did not compensate for the phase shift caused by the length of the cable. This time......." Yuri wondered.

There is a further board pattern from the cable tip to the end. Is it an effect of this?

Although the pattern length was short, the characteristic impedance was not matched to 50 ohms, so it was unsurprising that there was a certain amount of influence. The effect can be theoretically estimated if the affected pattern is simple and the characteristic impedance is known.

However, the actual pattern was made to put the matching elements in series and parallel, so it was impossible to estimate theoretically. The only way to find out was by actual measurement.

"Wait, well, if I can determine the influence by actual measurement, then I can take that into account to determine the value," she thought.

She tried to figure out the deviation of the arc obtained by actual measurement from the original theoretical arc.

No, it doesn't work when considering the arc as a whole.

She tried to imagine the position of each point in the original arc, assuming that each point was displaced from its original arc.

Yeah, I can kind of picture it.

It seems to offset from the theoretical original point on the arc in the direction in which the inductor was connected in series—not precisely, but roughly. However, knowing this does not immediately solve the matching value problem.

First, bring it to around 50 ohms and put a capacitor in parallel. This also lined up the points on an arc that deviated slightly from the original arc. Now, Yuri was no longer surprised. That's just the way it is. All she had to do was to find the matching values based on the actual measurements.

But it was a painstaking process. She would put up an inductor or capacitor with different values, measure it, figure out how much the value she needed, put it up again, and measure it. She repeated the process until she finally got a match. It took two days to determine the values.

Yet, she still just got the matching between the connector for the antenna and the frequency conversion IC. Beyond that, there were several places to get a match.

Oh dear, I'll just have to go steady.

Still, it gave her confidence to know that she could do it.

Yuri matched the other parts of the system: the output of the frequency conversion IC with the input of the IC that processes the IF stage and beyond, the IF output from the IF stage IC with the IF filter input, and the output of that filter with the input to the IF stage IC again.

Furthermore, the coupling between the frequency conversion IC and the local oscillator was loosely coupled, i.e., not tightly coupled, so the local oscillator side was not easily affected. Actual measurements also determined values around the discriminator used to demodulate the frequency-modulated signal. In the meantime, June was coming to an end.

Yuri had a strange feeling. In digital circuits, logic was everything. Of course, being electronic circuits, there were electrical problems, such as overshooting or undershooting of signal waveforms or timing problems, and logic was not the only problem. Nevertheless, the main problem with digital circuits remains the logic design. In other words, if the logic was correct, it worked. This was even more so because the digital circuits Yuri was working on were not that fast.

However, radio-frequency circuits are different. It does not go according to logic. Of course, radio-frequency circuits also operate according to the laws of physics. In that sense, they work according to logic. However, it is only possible to consider some of the physical laws in actual design. We design with a somewhat simplified, practical model. So, it does work the way that model does. And the model does not fully incorporate reality. With digital circuits, she was never aware that the model was a simplified version of the model. But working on radio-frequency circuits like this reminded her of this. Yuri was acutely aware that she was dealing with a physical phenomenon, which she also enjoyed.

Over the weekend, Yuri's father asked her.
"How's your job going? Are you starting to get the picture?"
"Yeah. I'm well aware that it doesn't work according to logic."
"Oh yeah. You're actually building circuits?"
"Yes."
"You're not going to use the simulator?"
"A veteran told me I should not use it until I feel the reality."
The veteran was Suzuki. He said that if you use a simulator from the beginning, you will not understand its limitations.
"Well, that's good. You should be thankful."
Her father looked happy but somewhat listless.
I guess, Dad, you're not happy that I became an engineer 'despite being a girl.'

Acceptance inspection

In July, there was an acceptance inspection at the customer for the product Yuri had been in charge of before moving to Development Section 1. This procedure was called "final inspection" by the customer.

Yuri was also a project member and was asked to participate. The project leader, Shibata, first gave an overview of the project and then demonstrated the operation of each inspection item according to the inspection manual. The explanation was given to the head of the client's department.

Yuri had also participated in the "preliminary inspection" three days earlier. At that time, the customer representative was the counterpart, and she had to join because technical questions would be asked, but not this day. This day, it was a ceremony, so to speak. Even if the department head had pointed something out, it could not have become a problem because it would have been simply "agreed to by the person in charge."

There is no need for me to be here, Yuri thought.

But really, the final inspection didn't matter. She just needed to be there. The problem for Yuri was that a reception was scheduled afterward.

As a woman, I have to go around amiably pouring beer; such a thought was depressing. Yuri was not good at such things.

But, well, I can't help it. I can't be affable, but I'll have to do it my way, Yuri thought. Then, she realized that the final inspection had ended without incident.

The acceptance inspection had been completed, and sales would be made. Yuri and the others were told to wait for a while in the room where the final inspection had been completed. After a while, the door opened. It was Takamura, the customer's technical leader.

He said, "Please go to the conference room, as we are ready for the reception."

When Yuri entered the conference room, canned beer, oolong tea, paper cups, and dry snacks were on the table. Seeing this, Yuri felt a little relieved. Even though it was a reception, it was just a light drink.

The reception began with Takamura's toast. Then, the chit-chat ensued. At first, the topic was technical, but gradually, it became about hobbies.

"Well, maybe it's about time," Yuri thought, and she decided to go around pouring beers before her hobby was mentioned. At that moment, an arm holding a can of beer suddenly reached out from her side. It was Takamura.

"Thank you very much, Ueda-san. Your response to our unreasonable request really helped us out."

"Uh, well......" Yuri couldn't speak further and just held out her paper cup.

It's not good to have the customer pour. I should be the one pouring for the customer.

Unaware of Yuri's feelings, Takamura continued to speak.

"We were really, really in trouble. Our salesperson told us that the product would not be viable unless we changed the specifications, and if we changed the specifications now, of course, we would not be able to meet the delivery date. But we decided to talk to Katakura-san, your sales representative, and within a few days, he told us that it was possible. That was really helpful. Thanks to that, we can ship this system by the end of the first half. Otherwise, our team's sales for the first half would have been zero. I heard that Ueda-san's idea made it possible. Truly, thank you very much!"

"Thank you for your words."

Yuri could only bow her head and say so.

Oh yeah, he is an engineer too. He is a system integrator who built systems using products from our company. It seemed the same for engineers everywhere to be asked to do too much by salespeople.

That was what Yuri was thinking when she noticed that Katakura, seated a short distance away from her, was looking at her and smiling.

He told Takamura-san that it was my idea.

Yuri nodded back in appreciation.

"I used to think an engineer's job was making things. But really, it's about making customers happy through things," Yuri felt deeply in her mind.

The wake

The next day, Yuri was back at work on radio-frequency circuits. No, it's not working; it's still more like studying.

Yuri was now running a PLL (Phase-Locked Loop) circuit, which can generate various frequencies by changing its settings. This circuit is indispensable for modern radio equipment. She experimented with different loop filter values to see how the characteristics would change.

Near noon, the phone at the department head's desk rang. Director Iwata answered the phone.

"This is Iwata. ...... Yes, I've read your email. ...... OK, bye."

Iwata put down the receiver and stood up.

"Gather around, everyone. It won't take long."

He raised his hands and motioned for everyone to gather around. Everyone took a break from their work and gathered around the director. What on earth could it be? Yuri had a bad feeling.

"Everyone, please calm down and listen. ...... Manager Yoshioka passed away this morning."

Everyone buzzed faintly. Yuri couldn't believe her ears.

Passed away? That's absurd. Were the rumors Ayaka said true?

The buzz quickly subsided. Everyone was ready to hear the rest of the story. It was an odd silence. She could almost hear her heartbeat. The silence was broken again by the loud voice of Director Iwata.

"I didn't tell you, but Manager Yoshioka had cancer."

Again, a brief buzz occurs.

He had cancer ...... after all, the rumors were true.

Yuri got angry. But she did not know what she was angry about.

"By the time he was admitted to the hospital in April, it was already too late," Iwata added words.

There was no longer a buzz. Only a few sighs could be heard.

"Then, I would like everyone to join me, preying in silence."

Everyone's back straightened.

"Silent prayer."

Silence prevails.

Yoshioka-san, thank you for everything. Please rest in peace. Yuri choked back tears.

"Please finish your prayer."

Iwata's voice is somewhat awkward. Still, the atmosphere in the workplace returns to normal, at least a little.

"The date, time, and location of the wake and funeral ceremony have not yet been determined. However, we may ask some people in Development Section 2 to help. I will email you as soon as I have more details. Well, that's all for now."

The buzz returned, and everyone went back to their desks. Yuri went to the front of the department head's desk and told Iwata.

"I will help, too."

Iwata looked at Yuri and said calmly.

"Well, OK, I will ask you, too."

"Yes."

Yuri bowed and returned to her desk.

Late in the afternoon, Yuri received an email. The wake—a ceremony held the night before a funeral—would be held two days later, on Sunday, and the funeral would be on Monday. The place and time of the ceremony were also written. It was said that they would ask for help only at the wake and that a few company employees would attend the funeral.

On Sunday night, Yuri was helping at the wake reception. The number of mourners was surprisingly large. The wake was at a funeral hall, but people were overflowing to the outside of the building. The rainy season had not yet ended, but fortunately, it was cloudy and not raining. Right next to Yuri, Shibata was guiding mourners.

"Please use the two reception counters on the right if you are company-related and the counter on the left if you are not. Once you have finished signing the book, please line up to offer incense."

Yuri was on the left side of the three reception counters, at the counter for non-company-related condolence visitors. Even though she was helping at reception, there wasn't much to do. The mourners wrote their names, left their incense money, and got in line to burn incense. All Yuri had to do was bow to the mourners.

Naturally, they were all people Yuri did not know. They were people around the same age as Yoshioka and high school students who seemed to be friends with Yoshioka's children. Even those who came to the reception counter for company relations were more likely to be strangers. She felt that all these people shared the same feeling. Sutra chanting echoed through the funeral hall.

"Oh, hi."

"Oh, long time no see."

Such small voices could be heard from time to time.

When Yuri raised her head after bowing dozens of times, she saw her father before her.

Dad!?

Yuri barely managed to keep her voice from coming out. Without saying a word, her father wrote down the name and placed the incense money taken out of the fukusa, a silk wrapping cloth. Then, he stood in a long line to burn incense.

Yuri's gaze was not fixed while her father waited in line to burn incense. Her father finished burning incense and went to the room of 'tsuya-burumai,' where dinner and drinks were served after a wake.

After a while, as the number of mourners decreased, Sato told Yuri.
"It's getting late, Ueda-san, so burn some incense and go home. I can't have a woman staying late."
Indeed, this funeral hall near Manager Yoshioka's home was a bit far from Yuri's.
"I understand. I will do so."

Yuri got in line for the shortened incense-burning line. At that moment, Yuri saw Yoshioka's photograph in front of her, which she could not see from the reception counter. As soon as Yuri saw the photo of the deceased, she almost cried aloud.

Don't!

Yuri put her hand over her mouth to keep her voice down. She could stop her voice but not her tears. Yuri hurriedly pulled out a handkerchief and wiped her tears; somehow, she held back more tears.

Yuri burned incense, had a few mouthfuls of 'tsuya-burumai,' the wake food, and then left the funeral hall.

"I'm home."
Yuri swept the purifying salt off her shoulders, and when she entered the house's front door, she found her father's shoes. She didn't usually see them when she came home, so it was something new to her. Yuri changed in her room and then went downstairs.

In the living room, her father was sitting on the edge of the sofa, his head in his hands. He was still wearing his white shirt on top but had removed his tie and changed into his regular chinos underneath. Yuri had never seen her father so depressed, and she felt even more depressed.

Yuri wanted to ask her father why he came to Yoshioka's wake and about the relationship between her father and Yoshioka. Still, she hesitated, wondering if she should talk to him.

Yuri entered the kitchen instead of the living room. She took out two glasses and filled them with ice. She reached for a bottle of whiskey in the back of the cupboard and noticed her mother standing in the kitchen doorway. Her mother was silent, but her expression said, "Are you serving liquor?" *Oh, right, Dad never used alcohol to drown his bad feelings*, Yuri recalled. Her father only drank alcohol when he had a good occasion or on some anniversary.

Yuri returned the whiskey bottle, took a plastic bottle of iced coffee from the refrigerator, and poured it into two glasses. She added milk and stirred gently.

"Dad, would you like a drink?" Saying this, Yuri placed one glass in front of Dad.

"I don't need liquor."

Dad said with his head in his hands.

"It's not the liquor."

Yuri then sat down at the other end of the sofa and sipped from her glass. A cold bitterness spread in her mouth. Dad lowered his hands and looked at the glass before him.

"Ah, thanks."

Dad sipped his iced coffee as if savoring it and breathed deeply.

When Yuri was about to speak to Dad, he put his glass on the table, got up, and took a CD from the shelf as if he had an idea. The jacket showed a photograph of a conductor and the words "Beethoven" and "5". Dad put the CD in the CD player, operated the remote control, and returned to the sofa. "Beethoven Symphony No. 5......" Yuri thought.

However, the melody through the speakers was not the famous opening 'da-da-da-dum' that Yuri had expected. It was a gentle melody of stringed instruments.

What?

Yuri cocked her head. Then, sitting between Yuri and Dad, Mom said, "It's the second movement."

Mom held a glass of orange juice.

"Yeah," Dad replied.

"Yoshioka-san loved it."

"He did."

Yuri got confused. How could Mom have known Manager Yoshioka, too? Seeing Mom and Dad listening to the music, Yuri could not speak to them.

The second movement was played by a small audio set but a slightly high-grade model. Thanks to the night mode, the sound did not become thin even at low volumes; the weakest parts could be heard well, and the strongest parts did not disturb neighbors.

Only the music echoed in the living room. Woodwinds joined strings. Brass and percussion instruments also joined in to add to the intensity. Then, back to the gentle melody again. It was a gentle but powerful music. To Yuri, this music seemed to represent Yoshioka's personality well.

When the second movement ended, Dad got up, took the CD out, put it in its case, and put it back on the shelf. Yuri wondered if she should speak to him as he sat on the sofa. Then Dad spoke up.

"Yoshioka and I were at the same college."

"What?...... Why didn't you tell me?"

"How could I say that? If you knew your boss and dad were friends, you'd have a hard time."

"Oh, I see."

"It must have been hard for Yoshioka-san," Mom added.

"Mom, how did you know him?"

"Because I met him when we got married. At the meeting for the wedding reception, or at the moving into company housing."

"Oh, right."

Yuri understood that it was only natural for a friend of Dad.

"Yoshioka-san and I had a lot of fun talking about classical music," Mom said.

"Hmm. Oh, so you talked about the second movement."

"Yes, then, Dad stubbornly said, 'I like the fourth movement.' Dad only listens to jazz."

Mom's tone is joking, but her face is sad.

"No, I was just joking."

Dad's voice sounded sad. Then, with one sigh, he continued.

"Yuri, there was no one that good. He was serious, sincere, and brilliant."

"Yes. And gentle."

"Yeah, that's right. He said you're good. He said you have potential."

"Really?"

"Really. You're brilliant, and you've got engineering sense and guts, that's what Yoshioka said."

Yuri was both happy and embarrassed.

"But, just......"

"What?"

"But just wish you would be more honest, he said."

Yuri was depressed.

"I'm not honest……"

Bingo. She had no words to respond.

"Oh, no, no. Uh. He said you have too much of an obsession with excellence."

"An obsession with excellence?"

Yuri had no idea what it meant.

"Accept that you're not good enough, it means."

"What?"

"I mean, if you don't know, you should just say you don't know."

Yuri felt like her mind had been stripped bare. Dad continued without caring.

"Because we all have things we don't understand. There's no shame in not understanding something."

"But under performance-based system……"

"No matter how much it's a performance-based system, it won't change that. Pretending to understand something you don't know won't get you anywhere."

Yuri recalled that Suzuki complimented her question. Dad's words, or Yoshioka's, sunk into her heart.

"I still don't understand a lot of things," Dad added.

The daughter turned to Dad with vigor, "Really?"

"It's true. Of course, I can design products and solve problems. Still, the engineer's job is to apply the laws of physics to create value, right? Physical laws and their applications are so deep that it's no wonder. Engineers are learning all their lives. We have to be humble about the facts. Besides, technology moves so fast, keeping up is hard."

Yuri was glad that Dad talked about the mindset of an engineer and that Dad said "we" instead of "I" or "you."

But what about his feelings for her, Yuri wondered. She dared to ask. She could ask now. Yes, it would be the only time to ask.

"Dad, you didn't want me to be an engineer, did you?"

"Why? I was glad you became an engineer."

"Really?"

"Yeah."

"But when I told you that I was praised for making digital circuits in the engineering club in middle school, you said, 'Despite being a girl....'"

"Did I?"

"Yes. You did. Didn't you want me to get good at cooking and sewing, like Mana?"

Then, Mom said with a start, "At that time, Dad said, 'Despite being a girl, you're amazing.' Didn't you hear him?"

Yuri hurriedly scanned her memory.

"All I heard was 'Despite being a girl,' I think."

"Dad, you're too shy to say it out loud," Mom looked at him and said.

He nodded twice, widely.

"Oh, I remember now. I said, 'Despite being a girl, you're amazing.' I was happy at the time."

"When Yuri told us you would apply to the University Department of Information and Electronics Engineering, Dad was pleased."

"And you, Mom?"

"I was happy, too, because Dad was so happy."

"I never knew that. .... I have always misunderstood it. I'm an idiot."

Yuri felt ashamed of her smallness. But somehow, she was able to expose it now. Mom grasped her hand and said.

"I'm sorry, Yuri. I should have talked to you properly."

Yuri felt Mom's hands were warm.

"No, no. Mana was preparing for the high school entrance exam, and Mom was worried about her, so it couldn't be helped."

"I'm really, really sorry."

"You don't have to apologize. I should have talked to you about it."

"Yuri, you are just like me. You should talk to someone about it when you're struggling, but you're too stubborn to do so."

"Mom? It's not true for you."

"It was. Until Dad helped me."

"Dad helped you?"

"I had a broken heart when I was young, and it was hard, but I hid that pain and tried to be strong. Then Dad said, 'Don't force yourself to be strong. It's OK to say it's hard.' It made me stop being so weirdly tough. It's not that I'd stopped but that I felt better. So ...... I fell in love with Dad, and we got married."

"I didn't know that."

Then, in a slightly louder voice, Dad said shyly and playfully, "I guess I'm the worst one who didn't make it clear after all."

"That's right."

Yuri and Mon mischievously joined their voices.

"Sorry."

Dad apologized in a playful tone and smiles returned to the three of them. May Yoshioka-san be smiling in heaven, too.

Business negotiation

Everyone came to work as usual on Monday. However, only Director Iwata attended the funeral. The division manager and the president were also said to attend.

Somehow, the atmosphere in the workplace was heavy. Everyone was feeling the same way. Still, they went on with their work. Yes, "life goes on nonetheless."

Yuri's feelings sank when she thought of Yoshioka. At the same time, she felt as if she had shed her mental armor. Moreover, she had the feeling that now she wouldn't feel a thing, even if a spear or an arrow pierced her heart.

But Yuri couldn't afford to stay downcast. She told herself she had to do her best because Yoshioka was counting on her. But she could not get anything done. Yuri's mind was like muddy water, unable to see the future. Then Manager Sato said, "Kirishima-san and Ueda-san, are you free tomorrow at ten o'clock?"

"Yes, I'm free."

"Me too, I'm free."

Kirishima and Yuri answered. Yuri was only studying in Development Section 1, so she was naturally free. And there have been no problems with the products she was in charge of in the Development Section 2.

"There is a technical meeting with Kamoshida Electric Industries for a new business deal, so please attend it. I'll have Ueda-san handle the wireless part."

What? I'm already in charge of a product?

But she was also uncomfortable with getting paid to study all the time. She couldn't say that she couldn't do it yet. Still, why her, not Kirishima?

"Is it me, not Kirishima-san?"

"Ah, yes. I have something else for Kirishima-san to do."

She wondered about the ambiguous reply, but it seemed best not to ask further.

"Kirishima-san, please support Ueda-san."

"Yes."

Kirishima responded to Sato's instructions and did not seem to have any doubts. Yuri was relieved to hear that Kirishima would support her.

"I understand. Where is the location?" Yuri asked.

"Conference Room 203."

Sato answered while operating the computer. He reserved the conference room.

"At our company? Not at the client?"

"Right. We'll have the customer coming in."

"Wow. That's unusual."

The next day's meeting was to hear Kamoshida Electric's request. This was a meeting to begin the review process, not to answer on the spot. From the Development Department, in addition to Yuri, Sato and Kirishima were present in Development Section 1, and only Shibata was present in Development Section 2. In addition, Inada, who is in charge of sales, was also present. Three people from Kamoshida Electric Industries came to the company: a sales representative, a technical representative, and a technical leader.

After entering the conference room and taking a seat after exchanging business cards, Uchiyama, the technical leader of Kamoshida Electric, said, "Ueda-san, you are a female radio engineer? That's very rare."

Manager Sato replied, "She is our young ace."

"Oh, that's reliable."

I'm an ace? No way.

When Yuri unintentionally looked at Manager Sato's face, Sato was smiling, but it wasn't clear whether it was a joke.

"Then can you tell us about the specifications your company would like to have?"

Salesman Inada urged them to get down to business.

After the meeting, Yuri returned to her office and told Sato.

"Please don't joke that I'm an ace."

"It's not a joke. It's true."

"How could I be an ace?"

"I pulled you out of Section 2 because you are an ace. Thanks to that, I faced strong opposition from Yoshioka-san."

Having Yoshioka's name mentioned, Yuri could not say anything back. Perhaps misunderstanding Yuri's feelings, Sato continued.

"When Uchiyama-san said 'female radio engineers are rare,' he didn't say it out of prejudice."

"Hm? I know. Why would you say that?"

"Well, if you know that, that's fine. I just thought perhaps you might have misunderstood."

These words reminded Yuri of how she had been feeling until recently.

"I am sorry. It is true that before, I would have misunderstood. But now, I don't misunderstand like that anymore."

Sato seemed slightly surprised at Yuri's change of heart but soon smiled.

"I see. That's good to know."

Then he added, "Kamoshida Electric even has more female engineers than we do. They even have female managers."

"Oh, I see."

Yuri felt embarrassment yet somehow nostalgia, looking back at herself until recently when she had been obsessed with trivial matters.

Product development

The product for which Yuri was in charge of the radio part was a wireless data collection device consisting of central and peripheral units. The central unit would be connected to the equipment developed at the client's site and connected by a network. At first, they would like to be provided with equipment for the experiment. In other words, it was going into mass production later. This was often the case with recent wireless devices.

Moreover, the frequencies used in this equipment had yet to be approved. The Ministry of Internal Affairs and Communications (equivalent to the FCC, Federal Communications Commission, in the United States) was currently deliberating on allowing this. They said the experiment would be conducted under an experimental station license. This was an advanced development in anticipation of the opening up of frequencies.

Regarding the wireless portion, Kamoshida Electric Industry's most important requirement was a small peripheral unit. The first step was to find a usable IC. Yuri searched the Internet and began to look for ICs.

When Yuri was searching the Internet, she received an email from Kirishima with information about a wireless IC. It's an IC called TXE3330 from Texas Electronics, with a PDF datasheet attached.

After a quick read, she found it was amazing. The radio circuitry and the CPU were on a single chip. Moreover, the data transmission speed was fast. Using only one of these, a peripheral unit could be created. Of course, antenna, battery, switches, LEDs, etc., were necessary. Yuri had yet to learn that such an IC existed. She knew that SoCs were used in cell phones, but this was probably the first IC in this wireless category. Looking at the supply status on the website, Yuri saw that mass production had just started. This was usable. Her sinking feeling was drowned out a little. The rainy season, which Yuri hates, has ended today.

After reading the datasheet, Yuri found that the peripheral unit could be made with a simple circuit using TXE3330. For the central unit, the board of the peripheral unit could be used as a wireless module.

However, the design effort is challenging even if the configuration is simple. At any rate, the required functions are contained in a single IC with many functions. And because it was highly functional, it was also complicated to use. If you design each function yourself, you may have to design it, but you do not have to understand how to use it. This is because planning is creating how to use it. But if you use such an IC, you must understand its functions and how to use it before you can even design with it.

Yuri carefully read the data sheet and application notes. It was a very well-made IC. The CPU load was light since the communication-related functions were implemented in hardware. Preamble detection, encryption, and error correction were also implemented in hardware.

Since this IC also had a built-in CPU, it was necessary to consider how to write and debug the program. In this respect, too, it had been well thought out. The program could be written to flash memory with two serial communication signals and debugged. The serial communication signals could be connected to a PC.

Adjustments had to be considered for radio because each unit had to satisfy the provisional radio standards. Yuri compared the setting items of the IC with the radio wave standard—only the transmission output power needed to be adjusted for this product. However, measurements were required on each unit to confirm that it satisfied the standard.

There were twenty peripheral units to be shipped and two central units. With 22 units, it seemed best to avoid manual adjustment and measurement. Manual adjustment would have been fine for two or three units, but many units would take time. In addition, it is easy to make mistakes because of the monotonous nature of the work. If a mistake is made in adjustment and measurement and a radio that does not meet the standard is shipped, it would be a big problem. It was better to have automatic adjustment and measurement.

Automatic measurement was relatively easy for Yuri. She just attached a GP-IB interface to the PC, sent commands from the program on the PC, and received the values. Yuri was very adept at making such a program.

The problem, however, was automatic adjustment. The TXE3330 settings needed to be changed from the computer, and that setting value must also have been stored in the peripheral unit. An EEPROM was required to store the setting values. Anyway, storing the peripheral unit's ID, etc., was necessary.

Now, how should the change of setting values be achieved? The software on the peripheral unit had to do this, which meant that the peripheral unit needed to interface with a PC. The TXE3330 had a built-in serial interface circuit, which was usable. All that was left was to put the antenna, switch, LEDs, and batteries on the board.

Shimojo was to be in charge of the software for this product. Yuri explained the concept to Shimojo and received his approval. Then, she drew the circuit diagram and started to design the circuit board pattern. For this product, she included more filter and trap circuits because the radio standards under consideration by ARIB, the Association of Radio

Industries and Businesses, will likely make the second harmonic standards more stringent.

By the time the board pattern design was completed and the board prototype was ordered, it was August. While waiting for the board prototype to be completed, Yuri decided on specific values for various settings. The past few days of consideration have calmed her sinking feelings. Just as muddy water becomes more transparent with time, Yuri's mind has regained a bit of positivity to look ahead.

The board prototype arrived on the morning of the day, about a week after the drawings were submitted for the board prototype. It was time for Yuri to mount the parts on the board. At this moment, something in Yuri's mind created a chemical reaction that changed her mind, as if muddy water had turned to fresh water instantly. Yuri soldered the components to the board. It is hot to do soldering work in the summer. Yuri's workplace was even hotter because the air conditioning was inefficient. Yuri tied her hair back and continued soldering.

Once the parts were soldered, the next step was to write the Texas Electronics sample program into the TXE3330's flash memory. The circuit Yuri designed had the same input/output port settings as this sample program, so it should work as is.

The first sample program to run is simple: an LED lights up when a switch is pressed. However, if it works, it confirms that the circuit is working. The important thing is to make sure that the program runs correctly on this board. Once you know the program works correctly, you can debug the board by modifying the program. If the program does not work in a circuit, it is impossible to check its operation with a program.

May it work!

Yuri's shoulders tightened. She turned on the power, pressed the switch, and lit the LED. When she released the switch, the LED turned off.

Yes!

Yuri clenched her fists. This was a relief. Her shoulders relaxed.

Next, another sample program was written. This time, it was to operate the radio section. The settings of the sample program, such as frequency, were different from those of the product, so the settings had been changed. The program was written, the spectrum analyzer was connected to the antenna terminal, and the power was turned on. Once the switch was pressed, the transmit output should have appeared on the Spectrum Analyzer screen.

Go!

Yuri pushed the switch, but the spectrum analyzer screen did not change. That's strange. She pushed the switch one more time, but it was the same. Yuri turned off the power. Her body tensed up.

Why is that? Hmm. What can I check?

To try, she turned up the frequency span on the spectrum analyzer to maximum and tried again. Then, the transmit output was displayed, but the frequency was way off.

"Huh? Did I set the wrong value?" Yuri's thoughts came out unintentionally.

When she re-examined the setting values, she found that some needed to be corrected. After correcting the setting values in the program, the transmission output was generated at the correct frequency. The tension was relieved.

Phew, thank goodness!

When she looked at the clock on the wall, it was almost time for the last bus. Moreover, she realized that Yuri was the only one at work. She hurriedly turned off the power, made sure the windows in her office were all closed, turned off the switchboard, and hurried to the last bus.

The radio section worked, and the impedance-matching process started the next day. Yuri then determined the inductance and capacitance values for matching, which she had done before using a network analyzer.

Measure the impedance, replace one inductor or capacitor with another value, measure the impedance, and repeat. This was done to match the impedance to 50 ohms. This was done at four locations: transmit output, receive input, antenna measurement terminal section, and antenna system. This process took about a week.

Once the matching was done, it was time to start finding the filter values. The purpose was to reduce spurious radiation, i.e., output power at frequencies other than those at which it should be transmitted. If this exceeds the specified value, the product cannot be shipped. The matching got off whenever Yuri changed the filter value, so she had to work with spurious radiation and matching. But it was not easy.

Ignoring the filter and taking the impedance matching first was a waste of time.

But it was too late to realize this now. Time cannot be turned back. Anyway, she had to work with both considerations. But summer vacation arrived before much progress could be made on that task.

Yuri was still studying radio frequency circuits at home during her summer vacation. Outside the window, the summer sun was dazzling. She was rereading the books she had already read, working hard to understand them again, not just take them for granted.

Yuri's father was also on summer vacation, and aside from weekends, it had been a while since the three of them, including her mother, had meals together.

"Yuri, aren't you going out to play this year?"

Dad asked Yuri while eating cold noodles for lunch.

"I'm not going. This year is special. I have to study radio frequency circuits."

"Hmm."

Dad said nothing more. Still, it seemed as if he understood how she felt.

After lunch, Yuri's father was listening to a CD in the living room. Yuri could also hear Cannonball Adderley's music in her room. While listening to it, Yuri continued her studies.

The clock was past three o'clock. Yuri went downstairs for a cup of coffee. Her father was dozing off in the living room, and the CD had long finished playing. Yuri poured a glass of iced coffee in the kitchen and walked into the living room with it in her hand. She put the Cannonball CD back in its case and put on another CD. Then her father woke up.

"Hmm? Michael Brecker?"

"Yeah. Yes, that's right."

"Yuri, do you prefer the tenor saxophone to the alto saxophone?"

"Yeah."

"Why?"

"Because .... Um, because it's big and cool."

"That's it?"

Yuri's father was stunned. Still, she didn't care.

"Yeah."

"Hmm."

It's been a long time since she's had a conversation like this with Dad.

After the summer vacation, Yuri resumed work on finding the filter's value. Because of the pattern's influence, it did not work as calculated. She just kept on cutting and trying for days.

Then, one day at lunchtime, Yuri looked at the calendar and remembered. It was August 21, her older sister Mana's birthday. Yuri took out her cell phone and sent a Happy Birthday message to her sister.

Happy birthday, Mana. How is the baby doing in your belly? I'm having a difficult time with the development of wireless products, which I'm not used to. But I am doing well, so there is no need to worry. Dad and Mom are doing well, too. I got to know Dad a little better last month for one reason. Well then, give my best regards to your darling.
From your cute sister, Yuri

It seemed out of character for Yuri, especially the last line. Most of the time, her emails were clerical, but she didn't mind. She was sure Mana would be pleased. Yuri sent the email.

A few minutes later, Yuri received an email reply from Mana. She was a little surprised, as she was not expecting an answer, but it made her happy.

Thanks for the email. The baby is doing well. I'm glad to hear that you and Dad were able to reconcile. We'll talk well when I return to your house for the birth.
From the one who loves her little sister so much

The last line brought tears to Yuri's eyes. She had never been told like that before.

"But what does it mean to 'reconcile'?" Yuri thought. Mana was not supposed to know about that night. Maybe Mom called her. Perhaps Mom had told that story over the phone that Yuri had misunderstood "Despite being a girl."

At that moment, a memory came back to Yuri. At that time, Dad told her, 'Despite being a girl,' and when she returned to her room in shock, Mana came in and said.

"Dad is happy. He said, 'Despite being a girl, you're amazing.'"

But at the time, Yuri did not listen. She had a complex about Mana, who was very affectionate and good at cooking and sewing.

"I don't want to hear you. Get out!"

Yuri said that to her sister. She had forgotten it for a long time. Or maybe she never tried to remember.

Yuri had never had a bad relationship with Mana. She just had a complex about her sister. Yuri was sure her sister knew this, so Mana said nothing more. The sister knew Yuri wouldn't listen to her even if she told Yuri about it.

Yuri typed a reply email.

Mana, I'm sorry for everything I've done. Thank you. I love you.
Yuri.

After looking at the typed message for a while, Yuri canceled sending it. She would tell her sister in person when they met. That would be better.

Yuri returned to work on the filter values, reducing spurious radiations and matching mismatched impedances. Filtering and matching were not easily reconciled, but the product could only be shipped if they reconciled.

If the impedance were not matched, the transmit output would be small, and the receive sensitivity would be poor. This would not meet the specification. If the impedances are not matched, much of the transmit output power will not reach the antenna and will be reflected back. Similarly, much of the signal power received by the antenna will not go into the receiver input and will be reflected back.

Suppose the strength of spurious emissions, i.e., signals at frequencies different from those originally intended to be transmitted, exceeds the specification. In that case, this would not satisfy the product specifications.

Yuri stared at the board, which did not have satisfactory characteristics.

"Oh, I was the same way," Yuri thought.

She didn't say what she thought because she thought it was useless. She did not listen to others. And then she said something she didn't have in mind. She was like that, just like the circuit board before her. The thought of it made her cherish the board.

But I'm not what I used to be anymore. I will get this kid right, too.

It was a difficult task, but she would fix the values, Yuri declared to herself.

"How are you doing? It seems kind of difficult," Kirishima called out to her.

"Oh, yes. It's a little difficult, but it's OK. I'll figure it out."

"Do you need help?"

"Thank you. But I'd like to try it myself. That way, I can learn more."

"OK. I see. But if you want to talk to me about it, just let me know."

"I will. Thank you."

The words seemed a bit unexpected to Kirishima. It was natural since Yuri had never spoken such words before. But Kirishima said nothing more and nodded, smiling.

By the time she had finished fixing the values, it was September. In the morning, Yuri got off the bus and walked to work.
"Morning."
It was Ayaka who tapped Yuri on the shoulder from behind.
"How is Kirishima-san?" Ayaka asked Yuri.
"How's what?"
"He's teaching you, right?"
"Yeah."
"Is he kind?"
Yuri thought about it for a while. Indeed, Kirishima was not too outspoken, but he was always there for her when she was in trouble. Yuri answered, "Well, yeah."
"Fell in love?"
"What's that? Why would you think so?"
"It's so suspicious," Ayaka teased.
"Don't be weird."
Yuri quickened her pace. She did not want to hear Ayaka say anything anymore. She did not want to think of Kirishima as a male consciously.

Trap

As Yuri was organizing the schematic and parts list, she received an email from Uchiyama of Kamoshida Electric Industries.

> We are sorry, but we would like to change the spurious radiation standard at the second harmonic frequency.
> The current specification is -50dBm, but please change it to -61dBm.
> Sorry for the sudden and significant change.
> The radio standard is about to be set at this value.
> Please consider this matter.
> If you have any questions, please do not hesitate to contact me.
> Regards,

"-61dBm!?" Yuri's eyes widened in surprise.
This was a severe value at the second harmonic frequency, where spurious radiation is most likely to occur. Yuri barely passed the current standard of -50 dBm. She sighed, leaned back in her chair, and looked at the ceiling.
"What's wrong?"
Kirishima looked at Yuri. Yuri raised her body and answered.
"The spurious radiation is changed to -61dBm. At the second harmonic. The email has also been sent to you, Kirishima-san."
Kirishima opened the email.
"This is a tough one. We may have to make a custom SAW filter. If the standard is set, it will eventually be released as a standard product, but we can't wait," Kirishima said after reading the email.
"Anyway, I shall try," Yuri said as if telling herself.

She considered it all afternoon, but achieving -61 dBm was still difficult. Impossible, she should say? But if the radio standard was going to be this value, it was an obvious request.

When she looked at the clock, the time was about seven o'clock.
I may go home early today and think things over at home.
Yuri was thinking that when Kirishima said.
"Hey, can't we use the self-resonant frequency of the capacitor?"
"I'm sorry?"
"Self-resonant frequency. As an equivalent circuit, a capacitor is a series circuit of an inductor and a capacitor, right? So, it has a resonant frequency. Above that frequency, it behaves as an inductor, not a capacitor."
"Oh, I remember seeing a graph of that. A graph with frequency on the horizontal axis and impedance on the vertical axis. The impedance goes down to a certain frequency and then up from there."
"Right. The frequency at which its impedance is minimal is the self-resonant frequency. I wonder if this could be used in a filter. I would call it a trap rather than a filter."
"I see! I'll try."

Yuri used software downloaded from the Internet to examine capacitor characteristics. She looked for a capacitor with a capacitance value whose self-resonant frequency is at the frequency of the second harmonic.

There was no exact match, but she chose the closest one and put it between the transmit output line and the ground. The second harmonic was then significantly reduced to -63 dBm.
"We did it! That's great!" Yuri was thrilled.
"It worked." Kirishima was also pleased.
"But the transmit output is also down."
"Yes. The impedance at the transmitting frequency also changes, so we'll

have to re-match the impedance. But we'll figure it out."

"Isn't it difficult to do both matching and filtering?"

Yuri recalled the hardship and became concerned.

"In this case, I think it is OK. Because even with impedance matching at the transmit frequency, the capacitor used as a trap is still short at the second harmonic," Kirishima said.

"I see. It sounds fine," Yuri said with relief.

It would take time to re-match on the transmit frequency. However, they wanted to get a feeling today that they could go this way. So, they added an inductor with a calculated matching value. In other words, a new inductor was added in parallel resonance at the transmit frequency with the capacitor for trapping.

The results were good enough as of today. The transmit output power was at the lower end of the permissible error range for one milliwatt of the specification, And the level of the second harmonic remained the same. If the matching was pursued, getting a margin for transmitting power was likely possible.

They looked at the clock on the wall and saw that the last bus was about to leave.

"Oops, it's getting late. Let's go home today," Kirishima said hurriedly.

"Yes, it took quite a long time," Yuri agreed.

"It's late, so why don't we have dinner and then go home?" Yuri said to Kirishima as they got off the bus in front of the station. Kirishima seemed a little surprised but agreed.

"Yeah, let's do that."

They entered a family restaurant in front of the station. For a late night, there were quite a few customers.

"Why don't we have a beer? Just a little. I want to celebrate."

When they got to the table, Yuri said, with a strange frankness even to herself.

"Eh... Well, a little bit, then."

Kirishima said, ordering a small bottle of beer, two glasses, and a hamburger steak set meal. Yuri ordered the hamburger set meal to match Kirishima's.

When the beer and hamburger set meals arrived, Yuri poured Kirishima and herself glasses of beer and said, "Well then, to the solution of the issue. Cheers!"

They clinked glasses, and Yuri downed her beer. The beer sinks into her throat.

"Ah, it's so good!"

Kirishima smiled at Yuri, who had a big smile on her face. His glass was still half full of beer. He drank it up and poured the beer left in the bottle into Yuri's glass, saying, "For your good job."

"Oh, thank you."

Yuri's glass was half full, and the beer bottle was empty. Yuri drank it up, and the two set to work on the hamburger set meals with each knife and fork.

"Kirishima-san, you are the opposite of me. You are excellent, but you don't flaunt it. I just pretend to be excellent."

For some reason, Yuri could say what she felt naturally.

"Hmm? You are excellent, Ueda-san."

Kirishima brought a hamburger steak to his mouth.

"No. But I'm not going to act anymore. I'll stop pretending to understand."

Yuri took a bite of rice.

"You didn't pretend to understand."

Yuri, with a full mouth, silently shook her head. Kirishima continued to eat, looking at Yuri with an incomprehensive look.

Finally swallowing her rice, Yuri said, "Anyway, I look forward to working with you from now on."

Yuri bowed her head.

Outside the train, which Yuri and Kirishima boarded, a station platform appeared in the window. The train was slowing down. Kirishima said.

"I'm changing trains here."

"Uh, yes."

The train stopped. Kirishima looked like he wanted to say something.

He said, "Actually, I'm probably going to leave the company. ...... but don't tell anyone yet."

"What?"

The doors opened. Kirishima got off the train with the flow of people. Yuri had no idea why. The doors closed. The train started moving. Kirishima's figure walking on the platform outside the window left her sight.

"Leave the company? Why?"

Yuri's thoughts stopped there.

The following day, Yuri wanted to ask Kirishima what was going on. However, it would be bad if others heard about it, and it would also be difficult for her to take Kirishima somewhere. Yuri sent an email to Kirishima.

I need your explanation.

Yuri paid attention to Kirishima, who was visible in the corner of her sight.

Kirishima seemed to have read the email. He started typing on the keyboard. Yuri received a reply.

I will give you the materials separately.

What's that?
As Yuri was thinking about what to do next, Kirishima told her, "This is the document you were talking about."
Kirishima held out a USB memory stick in his hand. Yuri silently accepted it.
What does it mean?
Yuri inserted the USB memory stick into her computer. A text file was inside, and she opened it.

Please don't email me about these stories. Our emails are being recorded. You know that.
I decided to leave the company because I wanted to be part of a more significant project. Besides, my company didn't train me well, so I taught myself to use radio frequency.
But after teaching you radio frequency, I realized I had to think and learn by myself. It may be my misunderstanding that the company did not train me properly. Maybe it was the way they taught me to learn on my own.
I might decide not to quit, but I already have a new job, and they want me to come soon, so I think I will probably leave.

When Yuri finished reading, she looked next to her and saw Kirishima was gone. Perhaps he was in the rest area. When Yuri went to the rest area, she found Kirishima drinking canned coffee. No one else was there.

"Once you've decided, haven't you? You've already decided where to change jobs and made a promise, haven't you? Then stop saying 'maybe' and keep your word!" Yuri said that much and left the rest area. Yuri didn't know why she was so angry.

Send-off party

Kirishima's send-off party was held last Friday night in September. The venue was an izakaya (Japanese-style pub) in front of the station closest to the company. Sato, Suzuki, and Sugiyama were having a great time discussing Denshi Block and MyKit. Yuri said as she poured Kirishima a beer.

"Thank you for everything."
"It was my pleasure to teach you, Ueda-san, and I learned a lot from you."
"Really? You also wrote in that file, is it true?"
"Yeah. I realized how difficult it is to teach others. I thought our company didn't train me properly, so I decided to change jobs partly because of that dissatisfaction, but maybe I was naive."
"Do you regret it?"
"No. That's not the only reason I'm changing jobs. I want to work on big systems at a big company. So, I think it's a good career change for me, but I feel bad for the others."
"You don't have to feel sorry. I never thought I would be able to design radio-frequency circuits."
"You never thought?"
"It's true. And maybe because of my move to Section 1, I can say these things."
"It makes me feel better to hear you say that."
Kirishima may have still thought that Yuri's words were a social call.
"It's true. Really, it's true," Yuri made sure of that.
"Kirishima-san, I wish you the best in your new job."
"Yes, I will. You too, Ueda-san."
"Yes."

"You can do anything."

"I don't think so. But I'm willing to try anything."

As she said this, Yuri spilled her beer on the table.

"Oh, no," Yuri hurriedly wiped the table with a hand towel.

"I did it again."

"Again? Ueda-san, are you surprisingly goofy?"

"Yes, I am. Even I am astonished at myself."

"Oh, at times like that, to Sato-san, I say, it pains me to admit this, ..."

Interrupting Kirishima, Yuri continued.

"But my naivete has cost us this mission.......right?"

Sato must have heard that, for he said, "Oh, Ueda-san, you know that? Char phrase? That's nice."

"Oh, did you hear me?"

Yuri turned to Sato and replied with a smile.

50 Causes of Arriving No Messages

Problem

As Yuri Ueda was preparing to leave, she noticed an email on her computer. She didn't have much time, but the subject line, "FW: Communication Error," caught her attention, and she opened the email. It was an email that a sales representative, Inada, forwarded to her from a customer.

>Shinomiya Foods has informed us of a communication error.
>Please consider this matter.

>>Occasionally, communication errors occur.
>>As far as I can see, messages do not arrive in two out of a hundred. >Is this frequency within the normal range?

(Hmmm ......)
As Yuri pondered while shutting down her computer, a senior colleague, Sugiyama, approached her.
"Ueda-san, let's go."
"Oh, yes."
Yuri followed Sugiyama and left the workplace.

Narumi Electronics Industry, where Yuri worked, was a medium-sized electronic equipment manufacturer. The company had two divisions: the Equipment Division, which manufactured and sold telecommunication-related equipment, and the Systems Division, which handled system integration.

For the first two years after joining the company, Yuri worked in Development Section 2 of the Equipment Division's Development Department, where she was in charge of digital circuit design. She then moved to Development Section 1, where she was in charge of radio-frequency circuit design for over a year.

Sugiyama was pushing a cart with cardboard boxes on it. They took the elevator down to the first floor.
"I'll get the key," Yuri said.
"Yes, please," Sugiyama replied.
Sugiyama pushed the cart toward the parking lot. Yuri headed for the General Affairs Department with an application form for using a company car in hand.

On the way, she passed her boss, Manager Sato, in the hallway.
"Ueda-san, are you on your way now?" Sato asked.
"Yes, I am," Yuri replied.
"May the glory of victory be yours!" Manager Sato then saluted Yuri. It was as if Yuri was the boss.
Here it comes. GUNDAM phrase, or rather, Char phrase. She wondered how to answer but couldn't think of anything.
"Sorry, I don't have time for this," Yuri said as she ran. In truth, she could have walked.
"Have a good day."
Yuri heard Sato's voice behind her. She could imagine him slowly waving his hand as he finished saluting.
"Yeah, I guess I failed this unannounced test," she thought. What would Kirishima have said?

When Yuri arrived at the General Affairs Department, she handed the application form to the staff.

"The key for car No. 2, please."

The staff stamped the application, bound it in a file, handed it to Yuri with the key, and said, "Then, have a safe drive."

"Thank you," Yuri received them and headed for the parking lot. The file also contained an ETC(Electronic Toll Collection system) and fueling cards.

When Yuri arrived at the parking lot, Sugiyama, who was waiting for her, received the key and loaded the cardboard boxes into the back of a company van. The cart was also folded and loaded. Sugiyama got behind the wheel, Yuri sat in the passenger seat, and they started the van.

The car carrying them entered the Metropolitan Expressway. The traffic was heavy, but the flow was good. At that moment, Yuri's cell phone rang. When she opened it, she found it displayed "Ayaka Nanase." Ayaka Nanase was a software engineer in the Systems Division Development Department and Yuri's peer.

"Hello."

"Can I speak to you now?"

"Yeah, fine."

"Have you read the email?"

"It's about the communication error, isn't it?"

"Yeah. What do you think, in terms of radio?"

"It's radio, so there's the possibility of errors."

"Then I don't have to do anything yet?"

"Uh, let's talk about it when I get back."

"I got it. So where are you going out today?"

"Giteki."

"Giteki. ...... Well, enjoy the drive."

"What are you talking about?"

Before Yuri could finish, the phone disconnected.

Ayaka, don't joke about driving. Moreover, her tone was still in business mode. I could forgive Ayaka if she were joking in her maiden mode, which she is good at. But if Kirishima was in the driver's seat instead of Sugiyama, ...

But soon, she reconsidered.

No, no, what am I thinking?

Yuri drowned out her thoughts. Even though he taught her radio for six months, she didn't have a special relationship with Kirishima. Besides, he had already left the company almost a year ago.

"What's about the communication error?" Sugiyama's voice brought Yuri back to herself.

"Oh, yes, it's the Shinomiya Foods system. I got an email before I left the office."

"What does it say?"

"Two times out of a hundred, they can't get the message."

"Two times out of a hundred, huh? ......"

"What do you think?"

"I wonder. ......"

An electronic display board showing the traffic jam status drifted backward overhead. There was not any severe traffic congestion ahead. Yuri and Sugiyama had plenty of time to spare, so they would arrive at their destination, TELEC, Telecom Engineering Center, well ahead of their test appointment time.

In wireless communication using radio waves, there are times when communication is not possible, even if the radio equipment is working correctly. One example is when a cell phone goes out of range.

Even if you are in a location where communication is possible, if a blockage moves in, the radio wave will be blocked, and you will not be able to communicate. Even if there is no obstruction, the reflection of radio waves may prevent communication. Because the length of the propagation path is different between radio waves arriving directly and those arriving by reflection, the phases of the arriving radio waves are also different. If the phases are reversed, they cancel each other out.

The system for Shinomiya Foods was installed on a food production line. Stickers printed with expiration dates and other information were affixed on the products in that line.

Occasionally, a printer malfunction or problem could result in erratic printing. In such cases, the image recognition device detected the abnormality, and the incorrectly printed items were removed from the line. At the same time, an alarm was issued.

Green, yellow, and red lamps indicated alarms. The green lamp lit up if normal, the yellow lamp lit up if one printing error was detected, and the red lamp lit up if two or more errors were detected within a specific time. When the red lamp was lit, an alarm sounded simultaneously. However, the factory was noisy, and alarms may not have been heard. Recently, due to streamlining, there have often been no workers nearby.

This was why the Narumi Electronics wireless system was introduced. The system used the signal that turned on the color lamps to transmit alarm messages wirelessly. A peripheral unit was connected to each image recognition device, and messages were sent to the repeaters throughout the plant. Each repeater was connected via a wired LAN to a host system — a

PC — from which a wireless LAN sent an alarm to the entire plant. A portable device carried by the worker received and informed the alarm. The Equipment Division developed the radio equipment for this system, while the Systems Division developed the software.

As she was being rocked in the car driven by Sugiyama, Yuri recalled when she witnessed the system's installation. It was a reasonably large and complicated old factory that had been repeatedly expanded, and the passageways were intricate. However, as a food factory, hygiene was taken very seriously. The floors and walls were stainless steel, free of dust, and the entrance and exits to the outside were disinfected by changing clothes, washing hands, and even the soles of shoes with chemicals.

In the system installation, peripheral units were connected to each image recognition device, and the repeaters were installed at the appropriate locations, mainly on the cable ladder.

A cable ladder is a ladder-shaped metal fitting suspended from the ceiling. Various cables run on the horizontally suspended cable ladder, which is used to run power and signal cables throughout the factory.

Since the cable ladder was placed high, it was easy to avoid obstacles to communication if the receiver was installed there. It was also convenient for routing power and LAN cables.

The peripheral unit was installed near the image recognition device, about a person's shoulder height. A person passing nearby will become an obstacle between the peripheral unit and the repeater. Although they have confirmed that such obstructions do not cause communication problems, metal shelves with casters also pass through the aisles in the factory. It would not be surprising if such obstacles prevent communication.

The problem was not only shielding by obstacles. If a radio wave was reflected by something, not between the peripheral unit and the repeater, a composite of the reflected wave and the direct wave was received. If the phases of the reflected and direct waves were opposite, they may cancel each other out, making reception impossible. Even if there were no problems at the installation time, the receiving situation would change if the reflection changed.

"There are many possible reasons for the communication failure, such as obstacles and reflections," Sugiyama mumbled as the car merged onto the Bayshore Route of the Metropolitan Expressway.
"Yes. There could be any number of possible causes."
"But that doesn't prove that the equipment isn't the cause."
"I agree."
"Perhaps the biggest problem is how to convince the customer," Yuri thought.

The car exited the Metropolitan Expressway at the Oi-minami Exit. The car turned right at the overpass intersection, made a U-turn at the intersection just after descending the overpass, entered the TELEC premises, and parked in the parking lot.

TELEC stands for Telecom Engineering Center and is one of the registered certification bodies stipulated by the Radio Law.

A radio station license is required to establish a radio station, such as a broadcasting station or an amateur radio station. However, users can legally use small-scale radio equipment such as cellular phones without obtaining a license, provided the manufacturer obtains giteki (technical qualification) or ninsho (certification) from a registered certification authority.

Sugiyama unloaded the cart from the car, loaded the cardboard boxes on it, and entered the TELEC building, pushing the cart. Yuri walked ahead of Sugiyama and pressed the elevator button. They got off at the second floor and went into the waiting room. They still had time before their appointment, so they had to wait.

Once Yuri sat on the sofa, she found a computer lined up against the wall with a sign that said, "Feel free to use it." Yuri used that computer and started browsing technical information sites on the Internet. Sugiyama had opened an industry newspaper placed in the waiting room.

As the appointment time approached, Yuri shut down the computer. Then Sugiyama stood up.
"A little early, but I think it's time to call in."
He picked up the telephone receiver next to the room entrance and pressed the dial button.
"Hello, this is Mr. Sugiyama of Narumi Electronics. ...... Yes, we will."
Sugiyama put down the phone.
"Test Room 3," he said.
"Yes."
They left the waiting room, and Yuri opened the door to Test Room 3. No one was inside yet. Sugiyama pushed the cart into the room, and Yuri entered and closed the door.

The room contained a computer, a table, a desk, and a rack with various measuring instruments built in. The computer and measuring instruments had been turned on, and the heat dissipation fan on the measuring instrument rack was humming.

Sugiyama took the central and peripheral units to undergo the giteki (technical qualification) test from a cardboard box and laid them on the table. The central and peripheral units were wrapped in air caps one at a time. Yuri opened her laptop on the desk next to the rack and turned it on.

The door opened, and the examiner, Akizuki, entered.
He said, "Good afternoon. Nice to meet you."
"Nice to meet you," Sugiyama and Yuri proceeded to make preparations again.

Sugiyama laid out the central and peripheral units on the table. Yuri plugged a serial cable into the laptop computer, attached a conversion connector to the coaxial cable extending from the measuring instrument to the desk, and connected the measurement cable she had brought.

"Today's test will be for specified low-power radio equipment, 20 NT500s and 5 NR500s, right?" Akizuki said, looking at the documents in the file in his hand.

"Yes, it is," Sugiyama answered.

NT500 was the peripheral unit, and NR500 was the central unit. Yuri connected the power cable to the stabilized power supply on the desk and started the measurement software. Akizuki said as he handed a piece of paper to Sugiyama.

"Well, for NT500, numbers 7 and 12, and for NR500, numbers 2 and 3, please."

In giteki testing, measurements are taken by sampling. It is not a 100% inspection. Regarding the number of units tested this time, two units would be taken out to be really tested for both the central and the peripheral units. The paper Sugiyama received has the number of extractions written on it.

Akizuki sat in front of the computer on the opposite side of the rack from Yuri's desk and said, "When you are ready, please send a continuous transmission from NT500, number 7."

Sugiyama used a Torx screwdriver to open NT500's No. 7 case and handed Yuri the circuit board inside. Yuri connected a coaxial, power, and serial cable to it, turned it on, and then used the measurement software to put the peripheral unit into continuous transmit mode, saying, "Yes, continuous transmission."

On the screen of the spectrum analyzer, a spectrum in the shape of Mt. Fuji. This was a familiar shape from in-house measurements.

When Akizuki confirmed it, he clicked the mouse. The spectrum analyzer's display disappeared. The signal's destination must have switched to another instrument, a power meter, or a frequency counter.

However, the spectrum analyzer display soon returned. The signal was switched to the spectrum analyzer again, and measurements were taken one after another, changing the measurement frequency.

Wow, it's fast. Our company's measuring instruments are ancient, which is probably why they are slow.

The measurement proceeded several times the speed of the spectrum analyzer Yuri used at work.

For Yuri, who was accustomed to in-house measurements, the measurement at the transmission was quick and easy.

"Now, continuous reception, please," Akizuki said.

Yuri terminated continuous transmission using the laptop's measurement software and set the peripheral unit to continuous reception mode.

"Yes, continuous reception."

Hearing Yuri's voice, Akizuki clicked the mouse.

Again, the spectrum analyzer measured at different frequencies. The measurements with continuous reception proceeded faster than those with continuous transmission.

"OK, then, number twelve, please continue to transmit."

Yuri terminated continuous reception and turned off the peripheral unit. She disconnected the coaxial, power, and serial cable from it.

When she turned around with the peripheral unit board in her hand, Sugiyama was holding the board of the number 12. She handed him the number 7 and received the number 12.

She connected the cables to number 12 as she had done with number 7, turned it on, and put it in continuous transmit mode.

"Yes."

The spectrum analyzer displayed a spectrum in the shape of Mt. Fuji. Akizuki clicked the mouse, and the measurement began.

"Huh?" When Yuri thought something was different, Mt. Fuji disappeared from the screen. There seemed to be a slight difference from the in-house measurement. But it was only for a moment, so she could not be sure.

"Continuous reception, please."

Prompted by Akizuki, Yuri put it in continuous reception mode. That measurement was soon over.

"Then, NR500 number 2, continuous transmission, please."

Yuri handed Sugiyama the peripheral unit number 12 board and received the central unit number 2 board. She connected the cables as she had with the peripheral unit, turned it on, and set it to continuous transmission mode.

"Yes."

Even before she said that, Yuri looked at the spectrum analyzer screen. After all, it was a little different from the spectrum of in-house measurements. Akizuki clicked the mouse, and the measurement proceeded.

"The peak wasn't there," Yuri thought.

The rest of the measurements were also completed without problems.

Akizuki confirmed the correspondence between the construction design documents and the components on the circuit board, the description of the rating nameplate, and that the case is held in place with Torx screws.

"That's fine. Thank you. Then I will bring you the seals and the certificate. And the data, too. Please wait a moment." With that, Akizuki left the room.

They had written on the application form that they wanted the data so they could get a copy of the measurement data. Although they could get it, it was for a fee.

"That was different from our in-house spectrum," Yuri told Sugiyama as she packed her cables and laptop computer.
"Really? Which part?"
"Our internal measurements showed a peak a little below the carrier. It wasn't there."
"Hmm."
"Well, either way, it's not a problem by any standard."
"It's a little bit of a concern."

In-house measurements showed a small peak around the base of Mt. Fuji. Since it was well below the upper limit of the radio standard, there was no problem in terms of the standard. However, if there was a difference between in-house and TELEC measurements, they were unsure if they could trust the in-house measurement data.

"Thank you for waiting," Akizuki returned.
"Here are the giteki seals, the certificate, and the invoice. Here is a copy of the data."
"Thank you," Sugiyama took them and handed Yuri the certificate, invoice, and a copy of the data. Yuri put them in her bag.

Sugiyama put a giteki seal on each of the peripheral units. Yuri put them on the central units. The product's serial number and the giteki number on the seal should match the corresponding table given to them with the seals. By affixing this giteki seal, the product can legally emit radio waves and be shipped.

After Yuri and Sugiyama finished applying the seals, they wrapped the central and peripheral units again in air caps. They placed them in a cardboard box and packed them with a laptop computer and cables.

"Thank you very much."
Yuri opened the door, and Sugiyama pushed a cart with cardboard boxes out of the testing room.

They took the elevator to the ground floor and walked through the lobby. They loaded their cars and left TELEC.

As the car entered the Metropolitan Expressway Bayshore Route from an interchange nearby, Yuri took out her cell phone and called her company.
"Hello, this is Narumi Electronics, Equipment Development Section 1."
It was the voice of Suzuki, an old-timer.
"This is Ueda. We have passed the giteki test. We're on our way back now."
"Roger that. Good work. Have a safe drive home."
"Yes. Bye."
Yuri closed her cell phone and put it back in her bag. Perhaps because she was relieved, she felt tired. She must have been nervous without realizing it.

When Yuri woke up, the car was almost to the office. She had fallen asleep before she knew it. She watched the cityscape drift by, relieved that she had completed the giteki process.

The company approached. The car entered the gate and stopped in the parking lot. They got out of the car and unloaded their luggage.

"I will return the key."

Yuri received the key and the file.

"Yeah."

Sugiyama pushed the cart.

Yuri dropped by the General Affairs Department. First, she printed out the ETC card usage record. She handed the record, the key, and the file to the staff.

"I return the key for car No. 2."

"Yes. Thank you."

Back at the office, Yuri put her bag on her desk, and Sugiyama arranged the central and peripheral units on the communal table.

Of the five central and 20 peripheral units obtained giteki, two central and five peripheral units were left, and the rest were to be shipped to the customer. Yuri also went to the table and helped pack the units for shipment.

Once packed, Yuri pasted a courier slip on the cardboard box and placed it at the pickup location. This was the end of the day for her. Yuri bought a can of coffee from a vending machine and took a break.

She thought, "The prototypes are now ready for shipment, and the next step is to conduct reliability tests on them and then ninsho of the mass production trial."

GITEKI in Japanese (technical compliance, in English) stands for 'GIjutsu kijun TEKIgo shomei' (technical standards conformity certification), and NINSHO (certification) stands for 'koji sekkei NINSHO' (construction design certification).

Both giteki and ninsho are the same in ensuring that the radio equipment meets radio standards.

In the case of giteki, only products brought in for inspection can be certified. In contrast, in the case of ninsho, products of the same design can be approved without limitation on the number of units.

However, in the case of ninsho, the characteristics of a single unit are tested, and the quality assurance system is examined. Instead, in the case of ninsho, the manufacturer is responsible for guaranteeing the characteristics of the certified model and can ship any number of units. But ninsho is more expensive. Naturally, obtaining ninsho for mass-produced products in large quantities is customary.

Yuri filed the deed and finished accounting for the bill. A copy of the measurement data received at TELEC remains.

"Come to think of it, ..." she thought.

At that moment, the chime for the end of the day rang, and she decided to think about the measurement data tomorrow. It was not urgent, and overtime work was only allowed with the division manager's permission.

The next day, Yuri spent the morning comparing the data measured at TELEC with the data from the in-house measuring equipment. However, all units measured at TELEC had been shipped out.

The TELEC data had a hard copy of the spectrum display, but the in-house data only had peak values, so it was impossible to compare the spectrum directly. Still, there were little individual differences, so comparing the spectrum with that of another individual should have been no problem.

Yuri compared the spectrum measured at TELEC for peripheral unit No. 7 with the spectrum measured again for peripheral unit No. 16.

As Yuri had expected, the measurement of the number 16 on the company's spectrum analyzer showed a small peak around the base of the Mt. Fuji-shaped spectrum, which was absent in the TELEC data.

"I wonder why ......" Yuri pondered as she looked at the screen of the Spectrum Analyzer.

Are radio waves coming from outside?

She tried wrapping the number 16 in aluminum foil. However, the spectrum did not change.

"Sugiyama-san, as I thought, they are different. The spectrum is different from that of TELEC," Yuri said.

Sugiyama looked into the spectrum analyzer screen.

"But it's not a problem by the radio standards, right?"

"Well, right, but don't you care about?"

"Hmmm, but I wonder if there's a way to find out why it's different."

"What's wrong?" Suzuki, who had somehow come right by, asked.

Yuri explained the situation to him.

"Hmmm." Suzuki crossed his arms and thought.

"Have you tried changing the transmit frequency?"

"I'll try."

The prototype that obtained giteki had a fixed frequency specification, but the frequency could be changed internally.

Yuri increased the transmission frequency using software on the computer connected to the peripheral unit. Since the spectrum's position on the spectrum analyzer screen shifted, the spectrum analyzer's measurement frequency also changed.

"It's the same," Suzuki said.

"Yes, it is," replied Yuri.

This means the peak frequency in question has risen by the same amount as the transmission frequency.

"How far away from the carrier is that peak frequency?" Suzuki asked Yuri. Yuri displayed the markers on the spectrum analyzer and measured the difference in frequency.

"10.7 MHz." Suzuki read the frequency.

"Yes," Yuri answered.

"That's a meaningful frequency."

"Hmm?"

"It's meaningful," said Sugiyama, who seemed to know what it meant.

Suzuki crossed his arms again and pondered.

"Well, let's put in an attenuator. The value is 10dB."

"Yes."

Yuri did as she was told and placed a 10dB attenuator at the spectrum analyzer's input connector.

"Yeah."

Looking at the spectrum analyzer screen, Suzuki nodded and said to Yuri and Sugiyama.

"How's it changed?"

Yuri read the signal strength from the screen.

"The carrier is down 10dB and …… What? The peak is down 20dB."

She put in a 10 dB attenuator, so all signals should be 10 dB lower than they were. Oddly, the peak was 20dB lower.

Suzuki asked Sugiyama, "What does it mean by that?"

Sugiyama seemed not to be able to answer.

Seeing this, Suzuki said, "It's IM, isn't it?"

"IM…… intermodulation, ah, yes! I see!"

"Intermodulation: ……" sounded familiar to Yuri, but she couldn't recall the content.

"Due to the nonlinear nature of the circuit, the sum or difference of the two input frequencies is also output," Sugiyama explained to Yuri.

"Strictly speaking, that means $nf_1 \pm mf_2$," Suzuki added.

Then, Yuri remembered.

"Oh, that thing where terms of the second degree or higher in a polynomial equation do bad things."

Suzuki and Sugiyama nodded.

"But there is only one input frequency in this case," Yuri asked.

"As an input to the measuring instrument, yes," Suzuki answered archly.

"Yes?"

"I think another frequency is getting in inside the instrument."

"Inside the instrument?"

"Yes," he said. "The frequency of 10.7 MHz is often used as an intermediate frequency in receivers, so it would not be surprising if it is also used inside the spectrum analyzer."

"Is there such a thing as intermodulation inside a measuring instrument?"

"Well, measuring instruments are machines and not perfect."

"Heh ...... I see. But how can you tell what caused it by putting in an attenuator?"

"Because you put in a 10 dB attenuator, but that peak was 20 dB lower."

"Well, ..., uh, you mean, each of the two frequencies is multiplied by 10 dB lower, so the result is 20 dB lower."

"That's what I meant. Well, it's just a guess. But it makes sense."

"Ah, the carrier goes down 10 dB with the attenuator, but does the 10.7 MHz signal also go down 10 dB?"

"If 10.7 MHz is the intermediate frequency, then the lower the input level, the lower it goes, you see?"

"Oh, yes, I see. So that's what you mean."

"I'm just guessing."

"Yes, it is. But it's hard to explain if we don't think so."

"Yes."

Even though the data is from a measuring instrument, it may not be correct.

A few days later, as Yuri summarized the results of a reliability test on a prototype, Sugiyama approached her.

"Ueda-san, Manager Sato is waiting for you in Conference Room B."

"Uh, yes."

She knew what it was about—to inform her of the assessment results. Yuri stopped the work and headed for Conference Room B.

"It should be an 'A' this year," she thought.

The assessment was on a five-point scale of S, A, B, C, and D.

"This is Ueda. Did you want to see me?"

Yuri knocked and opened the door to Conference Room B. Manager Sato sat alone in the conference room, which had one large table and chairs for eight people.

"Yes. Please come in," Sato replied.

When Yuri sat down across from Sato, he spoke to her.

"Ueda-san, when you joined the company, you were assigned to the Equipment Development Section 2 and in charge of digital circuit design. But suddenly, you were transferred to Section 1 and put in charge of radio-frequency circuits, which you were unfamiliar with."

"Yes."

It was truly hard for her. She thought she had done well, herself.

"You have worked hard and developed your skills."

Yuri nodded silently. To her, Manager Sato's voice sounded very clerical.

His voice was different from his usual friendly tone. However, Yuri would have been in trouble if he had said his usual GUNDAM phrases to her at this time.

"So, for my part, I would have given you an A, but after some adjustments throughout the department, I regret to say that you are now a B."

Sato handed the evaluation sheet to Yuri. The letter "B" jumped out at her, leaving Yuri speechless. She had worked hard and achieved many results, but still a "B"?

"As far as I'm concerned, you did a great job."

That was not the issue.

"I believe you deserve an A."

That's enough. Yuri didn't want to hear any consolation.

The final rating was a B. That was the company's conclusion. This was the place to notify her of the evaluation results. It was neither a consultation nor a discussion.

Even if she disagreed, she couldn't dispute it. Of course, she could have spoken out in protest, but that would not have changed the outcome. It could not even be reconsidered. No such procedure existed.

Yuri wondered how she could get an A on the final evaluation, but she didn't have the energy to ask.

"I understand."

"Please keep up the good work."

"Yes."

She thought she'd probably get another B next year if she kept up the work.

"And this is a different matter. ......" Sato added.

"Yes." What in the world, she wondered.

"I'd like you to mentor the incoming newcomer to our section."

"Oh, me?"

"Of course, I will ask Suzuki-san to support you, but I would like you to provide direct guidance."

"Can I do that? ......"

"Teaching others is the best way to learn, so don't worry."

"Yeah ......"

She didn't quite understand why it was OK, but she couldn't say no.

The following week, Yuri received an email from the company about a personnel promotion. Among them was Ayaka Nanase's name. She had been promoted. Ayaka must have gotten an A grade. Naturally, Yuri's name was not on the list because she had received a B rating. She didn't feel frustrated or envious. She just felt her body lose its strength.

The chime for lunchtime rang, and Yuri headed for the company cafeteria. She placed her B lunch pasta and salad on a tray and, as was her custom, looked for Ayaka. Ayaka was in her usual place. The seat next to her was empty. When Yuri sat down next to Ayaka, she remembered the promotion. She felt awkward, but keeping quiet about these things would make things even more uncomfortable. She took the plunge and spoke up on her own.

"Have you seen the company's notice?"

"What? I haven't."

"You got promoted. But I didn't make it."

"Hmm."

Ayaka didn't seem interested in promotions or assessments. She had always been like that. Yuri couldn't understand it, but Ayaka appeared to be that kind of person.

"You got ahead of me."

"I was just lucky. Next year, you'll get a promotion."

Promotions and evaluations were not determined by luck; there was no way to know what would happen next year. But in her way, Ayaka was concerned about Yuri, so she decided not to mention it.

"Yeah, I'll do my best."

"You will make it."

Yuri nodded silently. This was the end of the conversation about it.

Still, Ayaka was acting strangely today. She seemed to be thinking about something else, even though she should have been a little happy since she had been promoted.

"Hey, anything wrong?"

"What? Why?"

"Because you don't seem well."

Ayaka sighed and nodded.

"I broke up with my boyfriend."

"Whaaaaat!?"

"Don't be so surprised."

But Yuri could not help but be surprised. She didn't know much about it, but they must have been very much in love. Besides, Yuri couldn't believe that Ayaka would be dumped by her boyfriend. If Yuri were a man, he would never let her go.

"After all, she looks cute, has a good personality, and is solid. She's the opposite of me." Yuri thought.

However, that question was quickly answered by Ayaka's words.

"Well, instead of saying I 'broke up' with him, it's more accurate to say I'm thinking of breaking up with him," Ayaka said.

OK, so you want to break up by yourself, Yuri replied in her mind.

"I'm kind of tired of it."

What a reason.

"I'm kind of dumbfounded myself," Ayaka said. "But I can't help it, men and women. ......"

Well, it seems there is no choice but to leave her be.

Fortunately, Ayaka changed the subject as Yuri contemplated moving on to something else.

"By the way, what's up with the communication error? You said we didn't have to do anything over here yet."

"Inada-san, the sales representative, told the customer to wait, see how things go, and send us the logs if they can. After all, we haven't seen the phenomenon in our operational tests," Yuri explained.

"We'd be in trouble if we couldn't confirm that we could reproduce the phenomenon."

"I agree."

"If you get the log, I'll check it. If it's not sent to me, forward it to me."

"Can you check it?"

Right then, Ayaka was working on a different project than Yuri, so Yuri had no idea how busy Ayaka was. Yuri had thought Ayaka would not check the logs.

"Yes, because our software outputs the logs, we're used to seeing them."

"OK. Thanks."

On her way home that day, Yuri was with Ayaka on the bus. Since overtime work had been banned in principle due to deteriorating business performance caused by the recession, such an occurrence was no longer unusual. They got off the bus together in front of the station, and Ayaka pointed to the entrance of the station building and said, "I'm going to go to the CD store."

"Oh, well, I'll go with you," Yuri replied.

They walked into a CD store in the station building.

"Ayaka, it is unusual for you to come to a CD store. We usually download them."

"Yeah, but I buy CDs occasionally, too," she said. "I'm in the mood for some dark music."

"That's OK. You'd rather be thoroughly depressed so you can get back on your feet rather quickly," Yuri said as she looked around the store.

Then, her eyes caught sight of the classical section in the back of the store. She said, "I'm going over there."

"Yeah."

Ayaka then headed for the J-Pop section and Yuri for the classical section.

Yuri picked up a CD in the classical music section. It was Beethoven's Fifth Symphony, which the late Manager Yoshioka loved, especially the second movement. Her mother had a CD of this piece of music, and it was on a shelf in the living room, so she could listen to it anytime. Nevertheless, Yuri also bought a CD of this piece and listened to it occasionally. She also began to listen to Symphony Nos. 3, 6, and 7 and the four major piano sonatas.

Yuri had previously bought Symphony No. 5 on CD conducted by Seiji Ozawa. At that time, she chose it just for no reason. At this time, she had a Karajan CD in her hand. Somehow, she was also interested in Karajan. However, she hesitated to buy two CDs of the same piece, even though the conductors differed.

"Oh, Karajan. He's so cool."

While Yuri was unaware, Ayaka was next to her. She had already bought the CD she wanted and was carrying a bag from this store.

"Yuri, you listen to classical music?"

"Well, a little lately."

"I don't listen to much classical music, but I like 'Pathetique.'"

"Oh, yes. 'Pathetique' is good," Yuri agreed.

"Especially the second movement."

"Yes."

"The melody is like a fluttering butterfly. It's so cute," Ayaka got in maiden mode.

"It doesn't sound like a butterfly. It's not cute, either."

"What's it like then?" Ayaka went back to normal mode.

"Well, it's a sad ...... image of a sunset."

"Sad? And a sunset?"

"Ah, but that's the effect of the drama. ......"

"Drama?"

"What?"

Why can't we get through to each other? Yuri and Ayaka both twisted their heads. It was Ayaka who noticed first.

"Hey, are you talking about Beethoven's piano sonatas?"

"Sure."

"Oh, yeah. I get it."

"What?"

"I'm talking about Chayko."

"Chayko?"

"Tchaikovsky, Sixth Symphony, 'Pathetique.'"

"Oh, yeah. ......, that's the one."

That being said, Yuri thought she had heard a tune like that. But why that one?

"But we're talking about Beethoven now." Yuri said, pointing to the word "Beethoven" on the CD.

"No, no, we're talking about the symphony." Ayaka pointed to "Symphony."

"Eh, ..."

Yuri wondered why Ayaka was paying attention to that one.

Yuri was a little angry but reconsidered. There was no point in arguing over something like this, so she settled the matter.

"Oh, I see. ......"

"Well, I guess we're both right, and we're both wrong."

"I agree."

Yuri was not truly convinced, but she would not insist any longer.

Reproduction

Two weeks later, another email from Shinomiya Foods was forwarded to Yuri by Inada, the sales representative.

We received a communication error log from Shinomiya Foods.
Since the regularity of the problem is reported, the cause is likely to be on the part of our product, not reflections, etc.
I do not think we can leave it any longer.
We appreciate your consideration.

> We are sending you a log of communication errors.
> In this log, approximately one product is flowing every second.
> We intentionally made an abnormal one, in which the printed area is filled in black to make an image recognition error.
> While normal products flow in the line, abnormal products are intentionally mixed in about once every ten times.
> An error message is sent when there is an abnormal item, but once in sixty-four, the error message is not received.
> There seems to be no doubt that there are regularities.

"One in sixty-four?" Yuri focused on the number.
That's a nice round number. Sixty-four is two to the sixth power. For a computer that processes binary digits, it is like "exactly one hundred" for the decimal numbers that humans typically use.

The mere regularity of communication errors would lead one to suspect a product defect. Still, with such a nice rounded number, based on this circumstantial evidence alone, it was almost certain that the cause of the error was on the product side.

In this system, the logs were stored by the application software on the PC. The log information was saved as a CSV file that could be opened in Excel.

The log information was listed horizontally by time and message type, and the pairs were arranged vertically in chronological order. In other words, the more recent the message, the lower the line.

There were two types of messages: ERROR and ALIVE. ERROR meant image recognition error. However, if a message was sent only when an error occurred and was not received, it was unknown whether it was because no error occurred or the radio malfunctioned.

Therefore, the system was designed to periodically send a message despite no error. This message was called ALIVE because it meant the radio was "alive."

Yuri opened the CSV-type attachment file, and among the vertical rows of ALIVEs, ERRORs were mixed in at almost equal intervals. Looking at the time information, ERRORs were received approximately every ten seconds.

The email included CSV and XLS files, the latter of which was saved again in Excel. Yuri opened the latter and found that the original log information had been modified.

The cells where the ERRORs were received were colored. The number of the ERROR was appended to the cell to its right. As the number increased from top to bottom, it became one, two, three, and so on.

It was almost equally spaced up to 63. However, there was no ERROR around where 64 should come next, nor was there an ALIVE in place of ERROR. Only at that time was the received data completely missing.

The next place where ERROR was likely to be was actually ERROR. And it was numbered 1. Looking further down, it was still almost equally spaced up to 63, with 64 missing.

And the same went for the next one after that. No doubt. Regularly, one out of sixty-four times, ERROR was missing.

"This is a bug in the app," Yuri thought.
Given the phenomenon, it was natural to assume it was a bug in the app, i.e., the application software running on the PC.

For the radio, the history of the message in the past was irrelevant when sending or receiving a message. The radio only sends and receives the message at the time it was sent or received. So, it was inconceivable that the ERROR phenomenon exited once in sixty-four times due to the radio. If it were a random exit, it would be possible.

If the radio was not the cause, then the app was what remained. Yuri did not know the processing details on the app's side, but since it sent out warnings based on the status of past messages, there was no doubt that the history was involved. It was natural to assume that there was a bug in that process.

The app was the most suspicious cause. Ayaka created the app. Yuri picked up the receiver and called Ayaka's office.
"Hello, System Development Section," The person who answered the phone replied.
"Hello, this is Ueda from Equipment Development Section 1. May I speak to Nanase-san?"
"Nanase-san is out today and won't be back."
"Oh, I see. Well, that's fine then. Thank you. Goodbye."
Yuri put down the receiver. The email had also been sent to Ayaka, and Yuri thought she could talk to her tomorrow.

Late afternoon, the phone rang, and Yuri picked up the receiver.

"Hello, this is Narumi Electronics, Equipment Development Section 1."

"This is Nanase."

"Oh, this is Ueda. I got the log about the communication error."

"Yeah, I got the email."

"But you're out, right?"

"Yeah, I'm forwarding it to my cell phone. It seems like a bug in the app."

"I knew it. Do you think so?"

After saying that, Yuri thought she shouldn't have said, "I knew it," but Ayaka didn't seem to mind.

"Given this log, ..., well, I'll check it out tomorrow."

"OK."

"By the way, I've got some big news!"

"What?"

"I met Kirishima-san."

"Kirishima-san?"

"He's as cool as ever. I emailed you his card."

Yuri looked at her cell phone but did not receive an email. Then she looked at her computer screen and saw that she had received an email.

A file was attached to the email. When Yuri opened the file, she found a picture of Kirishima's business card.

"I did it."

Ayaka said so, but what did she mean by "I did it."?

"Hey," just as Yuri was about to say, the phone went dead.

Ayaka was a hard girl to deal with, and Yuri stared at the picture on Kirishima's business card.

"Ayaka said, 'I did it,' so Ayaka will probably break up with her current boyfriend and go out with Kirishima-san. Well, it's not my place to complain. It's none of my business," Yuri thought.

The next day, Yuri was trying to reproduce the problem. Usually, she should have waited for Ayaka to investigate, but since she had to make only a few changes to the test program she had already created, she decided to try.

Previous testing focused on ERROR to ensure the ERROR message reached the repeater unit from the peripheral unit. Yuri placed the peripheral and repeater units at her workplace about 20 meters apart and had the peripheral unit send only ERROR once per second. Still, the ERROR message was not received only about once every 10,000 times. She tried it with one ALIVE between ERROR and ERROR, but the result was the same. She presumed that the messages were not being received because of obstructions and reflections and concluded that this degree of failure was unavoidable. The radio environment changes even as people walk.

The difference between the customer's logs this time and the test was that there are many more ALIVE logs, with the occasional ERROR. They knew this would be the case in actual operation, but they intentionally used more ERROR in their testing to perform an accelerated test.

If an ERROR is produced once an hour in actual operation, then if you do a test that produces an ERROR once a second, you have tested the system 3600 times faster. A full day of testing would be calculated to have tested a decade of actual operation. The probability of a message not being delivered is so low that they figured that without this method, the test would take too long.

Yuri modified the test program with nine ALIVEs between ERROR and ERROR. The message interval was one second. In ten seconds, ALIVE would be sent nine times and ERROR once. If an ERROR is not sent once every 64 times, it should be once every 640 seconds. Considering this as one set, it would take less than two hours to complete ten sets of tests.

After Yuri started test operations, she returned to her original task: preparing the NT500 and NR500 for mass production prototyping.

She must first make arrangements for making the boards, i.e., parts lists and mounting instruction drawings. Once that is done, she must create a procedure manual for automatically measuring radio characteristics for outgoing inspection and an automatic measurement software operation manual.

The lunchtime chime rang. Yuri had planned to look at the test results before lunch but decided to do so in the afternoon, so she headed for the company cafeteria. She also wanted to hear about Ayaka's situation as soon as possible.

"I'm working on it. I'll check the log first thing in the afternoon," Ayaka said while eating her A-lunch. Ayaka's situation was the same as Yuri's.
"Anyway, we need to recreate the phenomenon."
Yuri nodded at Ayaka's words.
Then she casually asked.
"I heard you met Kirishima-san."
"Oh, yes, I did. Yesterday, at the exhibition. He looked good. He was concerned about Yuri."
"Really?"
"Yeah."
Even as she said this, Ayaka seemed not to hand Kirishima's business card

to Yuri. After all, Ayaka might be going out with Kirishima. In any case, Yuri decided not to talk about it anymore.

After the lunch break, Yuri looked at the log. The sixty-fourth ERROR was missing. And the sixty-fourth ERROR from there, too. It was just as the client had pointed out.
"This is a bug in the app, as I thought. I guess I can leave the rest to Ayaka," Yuri thought.

Yuri emailed the log to Ayaka and returned to work arranging the mass-production prototype. It was unexpected that the cause of the missing messages was a bug, not the radio environment, such as reflections or shielding.

A little later, Yuri received an email from Ayaka. It was the reply to Yuri's email.

I checked your log.
I got the same results here.
I will continue to look into it.

The next day, one new employee was assigned to Development Section 1. Until then, training had been company-wide, but from then on, it would be conducted at each workplace.
"I am Shin'ichi Kishimoto. Nice to meet you."
The newcomer greeted the section member at the morning meeting of the section.
"Then I will ask Ueda-san, beside him, to provide guidance," Manager Sato looked at Yuri and said.
"Yes," Yuri replied.

"Well, we have three days of department-wide training starting today, so on-the-job training will come after that."

"Yes."

Yuri and Kishimoto's voices came together. They couldn't help but laugh.

"Then, Kishimoto-san, please attend the department's training session now. It's in the conference room over there."

"Yes."

Kishimoto headed for the conference room. Then Suzuki said to Yuri.

"Ueda-san, look, don't teach him too much."

"What?"

"We have to let him think for himself."

"Yes. ......"

"Otherwise, he won't be able to learn it. It was the same with you, wasn't it?"

"Oh, I remembered......"

Indeed, when Kirishima taught Yuri, he would let her study independently first, and if she did not understand, she would ask him questions.

"Was that Suzuki-san's direction ......? But even if it was, it's ...." Yuri felt an emotion that she could not put into words.

No, that was not important. She needed to figure out how to teach him. But first, she must make arrangements for mass-production prototypes.

Analysis

The next afternoon, when all the arrangements had been made except for those related to the shipping inspection, Yuri received an email from Ayaka.

When there is no ERROR message, there is no signal from the repeater unit to the LAN.
Attached is the waveform of the serial signal to XPort.

"What?" Yuri's mind went blank for a moment.
She hastily opened the attachment. It was a Word document file with images of two waveforms. The first was a ten-second waveform, with a serial signal going up and down dozens of times at one-second intervals. The up-and-down pattern represented the message's bit pattern.

The familiar Yuri knew it was an ALIVE message without analyzing the bit pattern. However, it was missing in one place. At first glance, the ALIVE message seemed to be missing.

But the attachment also contained another waveform. This one was a waveform of 30 seconds, and the serial signal was going up and down dozens of times at one-second intervals.

There was one omission almost in the center. The message ten messages before it was circled in red and accompanied by the red text "ERROR message." The message ten messages later was similar. Therefore, what was missing was an ERROR message.

Since the serial data was missing, the cause was not on the application software but the repeater unit.

Serial data was sent by XPort to the PC as LAN signals. XPort is a single component whose shape is a larger LAN connector. Inside, however, is a computer. The XPort processes TCP/IP protocol information from the LAN outside the repeater unit and outputs it as serial data inside it. The serial data is received by the serial port of the CPU that controls the repeater unit. Conversely, serial data output by the CPU from the serial port is converted to TCP/IP protocol by XPort and output as LAN signals to the outside repeater unit.

With XPort, the CPU can connect to a LAN using complex TCP/IP protocols simply by using simple serial data.

The ERROR message did not appear not because the application was failing to take it. It was because there was no output from the repeater unit. Serial data was not being output from the CPU inside the repeater unit.
"Wow, the cause is on the repeater unit side," Yuri thought.

However, it was strange to Yuri. The repeater unit only sent and received messages each time; the previous history was irrelevant. There should have been no mechanism on the repeater unit that would lead to the cause of the message dropping out once every sixty-four times.

Nevertheless, the phenomenon clearly showed that the repeater unit was the cause. If so, it was natural to assume that the cause was firmware.

Firmware is a computer program that runs on the CPU of an embedded system. So, it is a type of software. Therefore, it is acceptable to say "software." However, an application program is also a type of software. So, if you say "software," it is hard to know whether it is firmware or an application program. Therefore, we distinguish between them by calling them "firmware" or "application."

This system's CPU program inside the repeater or peripheral unit was called firmware.

"The cause is not the radio environment or the application but the firmware. It's surprising," Yuri thought.

Yuri was not in charge of the firmware, so she did not know its internal details. She thought that the specifications should not have anything to do with history, but she thought there must be some internal process in the firmware that had something to do with history.

At that time, she received an email from Numajiri of Development Section 2. He was the person in charge of the firmware for the repeater unit.

The cause is on the repeater side.
If so, the firmware seems suspicious.
It is a bit unlikely from the contents, but I will look into it.

Well, it doesn't seem to relate to my area of responsibility anyway.

More importantly, Yuri had to mentor Kishimoto, the newcomer, starting next week. Yuri planned to have him read the book first, as she had done with herself, and then have him build and experiment with the actual circuit.

However, she did not know his level of expertise. Maybe he knew more than Yuri. On the other hand, she might have to teach him the basics.

Yuri started reading a book she intended to give to Kishimoto. The shipping inspection-related documents would not be needed for a while yet. The priority was to prepare for the instruction.

Yuri was reading the book at home on Saturday.
It wouldn't look good if I couldn't answer a question.
Reading the description of the Colpitts oscillator circuit, she recalled what Suzuki had explained to her last year.

Change the LC resonant circuit's C, or capacitor, to a circuit of two capacitors in series. The midpoint of the circuit is used as the reference voltage for the amplifier circuit. That was the key point. By grasping this, she got an idea of the principle of oscillation.

But here, Yuri had a question.

"Why a sine wave?"

She knew that if she set up a formula and solved it, it would be a sine wave. As Suzuki taught her, the essence of an oscillator circuit is a resonant circuit of an inductor and a capacitor, so if we formulate the equation, it is a second-order differential equation. Solving it yields the solution of the sine wave. She understood that much. But how could she imagine it being a sine wave? What does it mean to be a sine wave?

She wanted to ask Dad, but Dad and Mom were out for the afternoon. They had gone to a concert. Putting the sine wave aside for the moment, Yuri read the book.

After nine o'clock at night, Yuri was drinking coffee when Dad and Mom came home.

"We're home."

"Welcome home."

Dad was holding a bag from a convenience store. The contents appeared to be two bentos.

"I thought you had dinner outside."

"Well, it was crowded everywhere near the venue," Dad answered Yuri.

"Of course. Tokyo Dome can hold tens of thousands of people."

"Exactly."

Mom said in a tone that she had not noticed this before the event.

"Want some cold tea?" Yuri entered the kitchen.

"Yeah, thanks. Did you eat properly?"

"Yes, I did. Here you are. Tea."

Yuri placed two glasses of cold green tea on the dining room table.

"Thank you."

Dad and Mom began to eat their convenience store bento.

"How was it, the concert?"

Yuri asked, holding a glass of iced coffee in her hand.

"It was great. Anyway, I'm just glad I got to go to their concert. The first time they came to Japan, I couldn't get tickets, and the next time, I gave up from the beginning," Dad replied.

"I didn't know you liked Simon and Garfunkel," Mom responded.

"I'd barely listened to it since I started working for the company."

"Hm," Yuri said.

Naturally, Yuri had never heard of it either. To begin with, she had never heard of Simon & Garfunkel itself.

"And it sounded surprisingly good, too. Paul's guitar was great. And the singing, of course."

"There were solo songs, too."

"Yes, it was good to hear the solo songs. But I also wanted to hear '50 Ways to Leave Your Lover'."

"What's that?" Yuri asked involuntarily.

"Well, there is a song with that title. It means there are many ways to break up with your lover. I like the lyrics. ......"

Yuri felt Dad's story growing more passionate and decided to evacuate.

"I see. Well, I'm going to take a bath."

As Yuri left the living room, she could still hear Dad's voice behind her.

"The meaning of the lyrics is different at the end. The whole time, it's saying there are many ways to break up, like this way, this way, and then the way it

comes out at the end is an unexpected twist. ......"
Mom is being told a story. Yuri was almost forced into a long conversation.

"There are many ways to break up with your lover,......" Yuri thought as she soaked in the bathtub.
There are also many possible causes for not receiving the ERROR message ......
By the way, she had forgotten to ask about the sine wave.

The following Sunday, Yuri went downstairs from her room to the living room and found Dad listening to a CD of "Bridge over Troubled Water" while looking at the concert program. Yuri decided to wait for the CD to finish playing while drinking coffee. After the CD finished playing, Yuri spoke up as Dad put it back in its case.
"Dad, can I talk to you for a second?"
"Yeah."
"The oscillation waveform of an LC oscillator circuit or an LC resonance circuit is a sine wave, right?"
"Yes."
"If I formulate and solve an equation, I get a sine wave. I understand it. But I don't clearly understand what a sine wave means."
"Oh."
"How should I comprehend it?"
"Well, you know it will be a sine wave, right?"
"Yeah."
"So, what more do you want to know?"
"How can I say? I want to envision an image or ...... a way of seeing that makes sense?"

"OK. So you want to understand both in theory and by feeling."

"Yes, yes!"

"Mmm......"

Dad was pondering, staring at Yuri. Rather than thinking about the answer, he seemed to be thinking about Yuri, which was strange for Yuri. Dad finally opened his mouth.

"The way we feel is different from person to person. So you might as well think by yourself."

"By yourself? You mean ...... me? By myself?"

"Yeah."

"What?"

"What are you so surprised about?"

"Because I can't. No way."

"Why?"

"Because ......" Yuri could not put the reason into words.

"Maybe you think that kind of thing is something for someone great to do?"

"Yeah, because that's what they do."

"No, it's not true. For example, ... you have applied for patents, haven't you?"

"I have, but..."

"Then you can do it. It's all the same in that you must think of new things with your mind."

"But ......"

"But what?"

Yuri could not answer. She was not convinced, but she could not think of a counter-argument.

"Anyway, think by yourself. To be convinced by the senses, you must find the essence. And look at it from an angle from which you can see the essence."

"Essence......"

"Well, hang in there."

"Yeah. ......"

Yuri had no choice but to back down.

Kishimoto's on-the-job training began on the following Monday.

"First, read and study. Then, if you don't understand something, ask me a question." Yuri handed Kishimoto five books.

She thought about which book to choose but ultimately decided what Kirishima had given her. And she told him to read two of them first. The same two books that Kirishima had specified.

"Yes." Kishimoto received the book.

Kishimoto's seat was the one where Kirishima was. Yuri felt somewhat strange. Yuri turned to her desk.

Giving the books to Kishimoto would give her a few days — if he didn't ask her any questions — and she would have time to think. Yuri began to think about the sine wave case.

"You must find the essence," Yuri remembered Dad's words.

As Suzuki taught her last year, the essence of the LC oscillator circuit was the vibration of the LC resonant circuit. The essence of LC resonant circuits was ......, which Suzuki told her, too. It was the ratio of electrostatic energy to magnetic energy that kept changing. The sum of energy was constant. But what should she think beyond that?

"Hmmm, I'm at a loss. I'm already stuck," Yuri thought.

"I knew I couldn't figure this out alone," Yuri sighed, but she somehow managed to keep her thoughts moving.

Well, let's just try to make it an equation. I'll number the equations just in case.

$$E_E + E_M = E \qquad (1)$$

E_E is electrostatic energy, E_M is magnetic energy, and E is the sum of the two and a constant value. But she could not come up with anything after looking at this equation.

This is just a mathematical expression that the sum of electrostatic and magnetic energy is constant. The question is whether she can make any sense of this equation as a starting point. To do so, she would have to transform this in some way.

Wait, there was a formula for E_E and E_M. Yuri wrote down the formula. *I understand the formula, so I can use it as is.*

$$E_E = \frac{1}{2}Cv^2 \qquad (2)$$

$$E_M = \frac{1}{2}Li^2 \qquad (3)$$

C is the capacitance of the capacitor, v is the voltage, L is the inductance of the inductor, and i is the current.

Substitute equations (2) and (3) into equation (1).

$$\frac{1}{2}Cv^2 + \frac{1}{2}Li^2 = E \qquad (4)$$

Fractions are an eyesore. Let's double both sides.

$$Cv^2 + Li^2 = 2E \qquad (5)$$

Yeah, this is easier to see. Well, now ......

What can Yuri say from this equation? Can she find some meaning in it? Yuri looked at the equation.

In this equation, the variables are i and v, and the others are constants. Both variables are squared and are in the form of their sum. More precisely, it is a sum of constant multiples of squares.

That's the left side. The right side is a constant. Wait, ... *Pythagoras!* Yuri had a flash of inspiration.

Perhaps she could transform this into the form of the Pythagorean theorem. She tried to convert the equation.

$$\left(\sqrt{C}\cdot v\right)^2 + \left(\sqrt{L}\cdot i\right)^2 = \left(\sqrt{2E}\right)^2 \qquad (6)$$

OK, I got it transformed into the form of the Pythagorean theorem. Wait, I squared the square root. Is that OK? Let's see, ... Oh, it's OK. It's a positive real number in the square root, so it's no problem. OK, good.

Yuri made a diagram of the Pythagorean theorem.

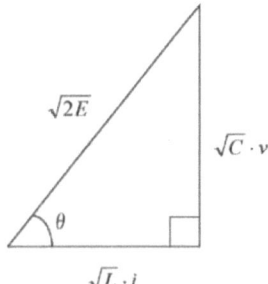

Well, what can I say about this diagram: ...... There are variables on two sides other than the hypotenuse: ...... Whoa, the hypotenuse is a constant! I figured it out!

Yuri added the angle θ to the diagram.

Now I'm getting sin and cos.

$$\sqrt{C} \cdot v = \sqrt{2E} \sin \theta \qquad (7)$$

By transforming equation (7), v can be obtained.

$$v = \sqrt{\frac{2E}{C}} \sin \theta \qquad (8)$$

I did it! v is now sinusoidal!

In the same way, she also obtained i from the figure.

$$i = \sqrt{\frac{2E}{L}} \cos\theta \qquad (9)$$

Huh? Are v and i constants, then? No, θ is a function of time.
So, she was not done until she found the equation for θ.
"Well, I can just differentiate equation (9)," Yuri had an idea.

$$\frac{di}{dt} = \sqrt{\frac{2E}{L}} \cdot -\sin\theta \cdot \frac{d\theta}{dt} \qquad (10)$$

Now that she had sinθ in equation (8), she could eliminate sinθ in equation (10) with equation (8).
Oops, the left side of equation 10 is also in the way. What to do?
Yuri held her head.

"I'm out of ideas? After all, the only way is to solve the differential equation?"
She stared into the air and thought.
She wanted to find θ differently, not by solving differential equations.
"So! I can use the definition of inductance!" Yuri got another idea.

$$v = -L\frac{di}{dt} \qquad (11)$$

She substituted equation (11) for v on the left side of equation (8) and then substituted equation (10). She then eliminated the two minus signs.

$$L\sqrt{\frac{2E}{L}} \cdot \sin\theta \cdot \frac{d\theta}{dt} = \sqrt{\frac{2E}{C}} \sin\theta \qquad (12)$$

She also crossed out sinθ and 2E on both sides and sorted them out. sinθ is non-zero, a condition she ignored. sinθ is a continuous function, even near zero, so it should be fine.

$$\frac{d\theta}{dt} = \sqrt{\frac{1}{LC}} \qquad (13)$$

OK, now θ is linear in t. Let's find it and let ϕ be the constant of integration.

$$\theta = \sqrt{\frac{1}{LC}} \cdot t + \phi \qquad (14)$$

I did it. I did it!
All that's left is to organize.
Let us do the following in equation (14).

$$\omega = \sqrt{\frac{1}{LC}} \quad , \phi = 0$$

Then, equations (8) and (9) become as follows:

$$v = \sqrt{\frac{2E}{C}} \sin \omega t \qquad (15)$$

$$i = \sqrt{\frac{2E}{L}} \cos \omega t \qquad (16)$$

Oops, E is still in the way.
Can't I use something to represent E? ...... Yes, for equation (15), let Vo be the maximum amplitude of the voltage.

$$v = V_o \sin \omega t \qquad (17)$$

In this way, since the current i is zero when the voltage v is at its maximum value Vo, the energy E is all electrostatic energy, which is determined by the voltage and the capacitance of the capacitor. It can be obtained by the formula just described.

$$E = \frac{1}{2} C V_o^2 \qquad (18)$$

She can also eliminate E here by substituting this into E in equation (16).

$$i = V_o\sqrt{\frac{C}{L}}\cos\omega t \qquad (19)$$

OK, now the formula for current i looks good, too.

That's all done. Let us summarize the concluding equations: equation (17), equation (19), and the definition of ω.

$$v = V_o\sin\omega t \qquad i = V_o\sqrt{\frac{C}{L}}\cos\omega t \qquad \omega = \sqrt{\frac{1}{LC}}$$

She reached this point from the formula that the sum of the energies is constant. She used the Pythagorean theorem, the electrostatic and magnetic energy formulas, and the definition of inductance.

After all, the essence is that the sum of energy is constant and that electrostatic and magnetic energy are orthogonal.

So, the angle θ changes at a constant angular velocity ω. Since the x and y components correspond to the current i and voltage v, they are cosine and sine.

"Wait, then I'd better have a different diagram than a triangle diagram," Yuri drew another diagram.

This figure removes the restriction on the range of angle θ. The angle with the x-axis is θ, and, of course, θ = ωt.

The distance from the origin is constant, corresponding to a constant sum of energy. This point, whose position is determined by the angle and distance, changes as the angle θ changes.

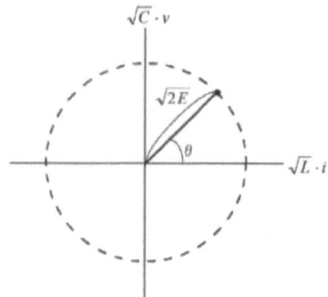

"It rotates around the origin!" Yuri thought.

The x-component of a rotating point corresponds to the current, and the y-component corresponds to the voltage. So, they correspond to cosine and sine, respectively.

Vibration is the rotation of energy!

So, when you look at the voltage and current components of the rotation, it appears as a sine wave. That is natural. They are the x and y components of rotation or circular motion.

Yuri felt like she was looking at a landscape she had never seen before. No, it was more than a landscape. It was more like a view from God's point of view.

"Hmmm, I see. ......" She couldn't help but voice her feelings.

Then Sugiyama heard this and spoke to her.

"Excuse me. Ueda-san, you've taken it very seriously since a while ago."

"Oh, I was just trying to figure out why the oscillation waveform is sinusoidal."

"Eh? But that's......"

"Well, I know that if you solve the equation, you get a sine wave. I was just trying to figure out how to capture that image or what it means."

"What it means. Hmm... Describing it in a formula is important, but I wonder if trying to find meaning is a good idea."

"Why?"

"Because that would be more philosophy or religion than science, wouldn't it?"

"Is that so?"

"I sort of think so."

"Suzuki-san, what do you think?" Yuri asked Suzuki, who was listening with interest, for help. Suzuki would understand.

"Well, I can see both sides of the argument, or maybe it just depends."

"How about in this case?"

"I have never thought about the meaning of a sinusoidal oscillation waveform. That's my first impression."

"Oh, that's the same for me," Sugiyama agreed.

"What do you think, Kishimoto-san?" Suzuki turned to Kishimoto.

"Is it me?"

"Yeah."

"Ah, I never thought about what it meant to be a sine wave."

"What do you think about seeking meaning in mathematical formulas?"

"Well......"

Kishimoto pondered. He had just been assigned to the company, so he was probably being reserved. Instead of Kishimoto, Sugiyama opened his mouth.

"Suzuki-san, it's impossible. He still doesn't know who is scarier, me or Ueda-san."

"Oh, I see," Suzuki agreed.

"Wait a minute. I'm not scary," Yuri protested with a straight face.

"Aren't you?" Sugiyama teased.

These words reminded Yuri of herself until a year ago.

"Haven't I been a little gentler lately?" Yuri managed to mend the situation.

"Ah, well, yes," Sugiyama seemed puzzled by Yuri's serious answer.

Sheesh, I should have returned with a joke.

"So, did you find out something? Ueda-san," Suzuki changed the subject.

Thank God.

"Yes."

"OK, let's hear it."

Yuri explained the contents at the meeting table while writing the equation and diagram on the whiteboard. She had just finished thinking about it, so she explained by looking at her note.

"So, since oscillation is a rotation of energy, we see a sine wave when we look at the voltage and current components."

"Oh, I see. Rotating energy," Suzuki responded.

"I think it's possible to ask for meaning in this. What do you think?" Suzuki asked Sugiyama for his opinion.

"Yes, I think it's possible because it's not so much that you're looking for meaning in the equation but looking at it from a different angle and formulating it from that point of view."

Hearing this, Yuri was relieved.

"Um, ..."

It was Kishimoto who said this in a reserved manner. He pointed to the last figure.

"Doesn't this look like a complex plane?"

"Wow!"

"Oh."

"Hmm."

Yuri, Suzuki, and Sugiyama all made three different noises.

In AC theory, complex numbers simplify calculations when dealing with sinusoidal signals. Yuri's diagram closely resembled that illustration.

"It kind of feels that way."

"Yeah, it has a vibe."

"It's got a vibe, but I'm unsure if it's strictly true."

Hearing Sugiyama's doubts, Yuri reconsidered.

"When looked at strictly, first of all, the dimensions of the quantities represented by the x- and y-axes are different from those of the complex plane."

"I think so. First, I want to remove the constant multiplier and set it to v and i."

When Suzuki said this, Kishimoto continued.

"If you make them v and i, it's not a circle, right? That would make the energy not constant, right?"

Sugiyama answered that question.

"If you change the dimension of the axes to v and i, the distance from the origin is no longer energy. In the original diagram, the dimension of the axes is energy, strictly the square root of twice the energy."

"Well, yes."

Yuri had to agree with this.

"It's difficult. If it corresponds to the complex plane, then the imaginary part would make sense."

"Usually, when we calculate complex numbers, we use their real part, not their imaginary part. After all, does it not correspond to the complex plane?" Yuri thought.

"Well, this is homework," Suzuki said.

"Whose homework is it?"

Yuri felt she would be unable to answer if it was her homework.

"Well, homework for all of us. Anyway, we have to get back to work. Manager Sato is glaring at us."

Suzuki said so, and Yuri looked at the manager, who was indeed looking at them.

"No, no, no, I'm not glaring at you. I was just thinking that you've gotten better," Sato replied.

"Yes?"

"Uh, forget it."

"Well, let's break up before we get more glares."

At Suzuki's words, the four returned to their desks. Yuri also returned to her desk, deactivated the screen saver on her computer, and entered her password. Yuri was about to return to work, but something stuck in her mind—Sato's words.

"You've gotten better ...... this is a ...... Char phrase!" Yuri had a flash of inspiration. She looked at Manager Sato, but he did not seem to have changed and was still working.

"I wonder if Sato-san is missing Kirishima-san ...... but it's a too high level for me. How would Kirishima-san say ...... No, I'll stop thinking about that," Yuri drowned out the thought.

Yuri decided to start working on the paperwork related to the shipping inspection. She could not put it off too long, so homework would have to wait.

She ran the automated measurement software and took screen copies of each step. Once that was done, Yuri started the word processing software and pasted the screen copies in order. She then described each step, mentioned the automatically stored logs, and instructed them to make backups. Now, the procedure manual was OK. The rest was the process chart created by Development Section 2, so she could send them the radio-related information and have them write it down based on that.

The cause

When arrangements for mass production prototypes were made, and preparations for certification began, Kishimoto spoke to Yuri.

"Ueda-san, I've read two of the books."

"How is it? Did you understand them?"

"Yes, roughly."

"Hmm...... 'roughly' .... What about the Smith chart? Do you understand?"

"Yes, roughly."

"Well, How roughly? What degree?"

"Well ...... what can I say ......"

"Then try to explain the Smith chart briefly."

"Yes. ...... Smith chart is a plot of the reflection coefficient from the impedance."

Kishimoto picked up the book, turned the pages, and said, "This formula calculates the reflection coefficient."

Well, I wonder if he understands or not, ... Yuri was unsure how to respond. She could have explained it immediately, but she needed to find out if Kishimoto would accept it. It would be nice if he had doubts, but if he did not, he would not be convinced by the explanation with a real sense of reality. Yuri decided to ask a few more questions.

"Aside from the equation, what is the relationship between reflection coefficient and impedance?"

"Aside from the equation?"

"Yes. Try explaining it without using formulas."

"A formula defines it, so how can I ......"

Just as Yuri thought, Kishimoto did not understand. The main problem was that he probably didn't know what he didn't understand.

"Then do you know why we use these things?"

"Why? I don't think there was anything in the book about why."

"Well, try to think about it."

"Me?"

"Yes. You figure it out."

"Yes. ......"

Yuri remembered her father telling her to think for herself. She felt weird and wondered if she was at a level where she could say something like this.

"Kishimoto-san, I don't mean this in a mean way," Yuri couldn't help but remind him.

"I know, I know," Kishimoto nodded.

Leaving Kishimoto to think about it, Yuri turned to her computer to find an email from Numajiri.

> I found a little bit of the cause of the problem.
> The interrupt of receiving data comes in twice with a short time interval.
> Before the main routine processes the data read by the first interrupt, the second interrupt comes in and overwrites it, resulting in no data.
> This behavior occurs because the same interrupts are not expected to appear at short intervals.
> This can be handled by making the data buffer a ring buffer, but the specification does not allow interrupts at such short intervals.
> One possibility is that it is receiving packets destined for other repeater units.
> However, if the address is for another repeater, the packet will not be received; if the address settings are incorrect, they will receive more

packets.

Besides, the received data for the second interrupt is "no data."

In other words, the problem is caused by two interrupts, but we do not yet know the cause of the two interrupts.

The phenomenon occurs once in sixty-four times, but there are cases where the phenomenon does not happen (does normally receive) at the time when it should happen.

Reading this, Yuri felt bad. It was strange to have two interrupts, and moreover, they were normal in some cases.

What does it mean? What's going on? Apparently, it has something to do with the hardware. I really need to look into this.

When Yuri thought this, Sugiyama called out to her.

"Ueda-san, I got an email about a communication error, but doesn't it sound like it's caused by hardware?"

"Yes, I think so. I'll look into it in detail."

"First of all, we need to look at the interrupt signal on the oscilloscope."

"Yes, I will."

Yuri left her desk and moved to an experimental table next to it. A pair of repeater/peripheral units with a wired connection, an experimental computer, and measuring instruments were placed there.

She operated the radios under operational conditions and observed the repeater unit's interrupt signal with an oscilloscope.

This interrupt signal was output from the radio module to the CPU, and its voltage changed from high to low when a packet was received.

When a "falling edge" of the interrupt signal was input to the CPU, the firmware — the program inside the CPU — would interrupt the processing and move to the interrupt routine. The interrupt routine read information on received packets from the radio module. A serial interface was used to read the packets. Once the packets were read, the firmware operation resumed the interrupted process.

The firmware periodically checked for the presence of received packets. Once that check was performed and the packet was found to be present, processing was performed based on the data in the packet.
To be specific, the LAN interface output ALIVE or ERROR messages.

Observing the interrupt signal with an oscilloscope was relatively easy. The problem, however, was timing. It was necessary to catch the waveform at the exact moment when the ERROR message exits.

Fortunately, in the test environment, the signal to the peripheral unit was made by a program on the computer. Yuri was able to modify the program to output a signal when the ERROR message did not appear. She could have put that signal into the oscilloscope and triggered it with that signal.

She could do this in this case because she knew when the ERROR message sent by the peripheral unit would not be output by the repeater unit. The oscilloscope waited until it was triggered and then displayed the signal when it was triggered. Yuri modified the program for testing so that the signal for triggering was output.

Under operational conditions, she connected the repeater unit to the peripheral unit. She set the oscilloscope to single-shot trigger mode so that subsequent triggers would be ignored once the first trigger was fired.

Yuri soldered a lead to the interrupt signal and pinched its end with a probe. Then, she started the oscilloscope and ran the test program. Now, all she had to do was wait until the oscilloscope was triggered. Until then, she had to wait for 640 seconds, or 10 minutes and 40 seconds.

The monitor software displayed a message on the screen of the computer connected to the repeater unit every second: ALIVE nine times, ERROR once, and so on.

Yuri considered making the interval between messages shorter but decided to leave them at one-second intervals because changing several conditions simultaneously could lead to unexpected phenomena.

The monitoring software displayed the time of receipt before the message; the ERROR message comes once every ten seconds, so she only needed to look at 10 minutes and 40 seconds after the start time. Of course, even if she did not look at the time, she could tell by looking at the oscilloscope as it triggered.

Yuri waited patiently. The passage of time seemed to slow down, but still, it was getting closer. She began to feel somewhat nervous.

"Come on, here it comes!"

The oscilloscope was triggered, and the waveform of the interrupt signal was displayed. Immediately, Yuri looked at the computer screen.

"Hm? It's not missing."

The ERROR message was successfully received on the screen. Numajiri said in his email that sometimes the phenomenon does not appear. That must have been what this was. She decided to wait and see for a while.

Yuri watched for about an hour, and the ERROR message always appeared fine. There was never any abnormal behavior.

"What's going on?" She was becoming confused.

"How's the situation?"

It was Sugiyama who called out.

"I don't see the phenomenon."

"Not at all?"

"No. Not even once. There were already five times when it was supposed to happen."

"Maybe it's the oscilloscope. ...... Take the oscilloscope off. And the lead."
"Yes."

Once Yuri stopped the operation and turned off the power, she removed the oscilloscope probe and the soldered lead as instructed by Sugiyama. Then she turned it on again to start the operation. Now, she had to wait another 10 minutes and 40 seconds.

When that timing came, this time, the ERROR message was dropped. The ERROR message was also dropped the next time and the one after. The abnormal behavior was reproduced.

"So it's the oscilloscope," Sugiyama said. "We have to use an active probe."
"Oh, yes. I understand."

When measuring waveforms with an oscilloscope, a probe extracts signals by contacting the circuit to be measured. This probe is usually passive. However, touching the circuit with a probe will affect it. The effect is generally small and is not a problem with passive probes.

Sugiyama advised using an active probe, which has less effect on the circuitry, because the passive probe's impact made a difference in this case.

Yuri switched the passive probe to an active probe and started over. When observing a specific signal with an oscilloscope for a long period, as in this case, it is common to solder a lead wire to the signal and pick up the end of the lead wire with a probe, as she had been doing. This is because it frees up her hands and makes observation work easier.

But if she did that when using the active probe, the leads' effect might defeat its purpose. This time, she held the active probe in her hand and observed the signal.

When she brought the hand-held probe into contact with the board, there was a risk of accidentally shorting the circuit. So she first applied the probe with her left hand with the power off, then turned it on with her right hand to start the operation. Then, she also started the oscilloscope in single-shot mode.

She had to wait more than ten minutes with the probe in her left hand. She must not relax; exerting too much force would also be bad. It is much harder than it looks from the other side.

I hope 10 minutes will pass soon.

Time seemed to pass more slowly to her than during the first measurement.

Three minutes. I have a long way to go.

She kept holding the probe still on the circuit board.

Five minutes. I'm only halfway there.

She began to lose strength in her hands. She regained her composure and just waited.

OK, I'm almost there.

As she waited, watching the oscilloscope screen, the oscilloscope was triggered, and the signal waveform was displayed. When she looked at the computer screen, the ERROR message was missing.

I did it. I got the phenomenon.

Yuri removed the active probe from the board. She looked at the oscilloscope's waveform as she relaxed her stiff-muscled arm.

"Hm?"

The waveform was just a little bit weird. As a whole, it had a decent high-to-low signal drop. However, there was a slight wave on the way down. There was a slight increase in voltage in the middle of the voltage drop. After that, the voltage dropped again.

Is this the cause?

She saved the waveform and then applied the active probe again to the board, which was still operational. Yuri did this carefully so as not to short-circuit the board with the probe. The oscilloscope was then started. She clicked a button on the PC screen. This would cause the signal for triggering to be output at the timing of all ERROR messages.

After waiting a few seconds, the oscilloscope triggered, and a new waveform appeared. Overall, the waveform was similar but not wavy. The voltage was decreasing monotonically; the ERROR message was not missing. *Just as I thought, it's wavy, so the interrupt comes in twice. ...... But why is the waveform wavy?* Yuri thought as she saved the waveform.

Since the investigation of the phenomenon's cause could not proceed after only one observation, Yuri made the same observation several more times. In each case, the same phenomenon was reproduced.

Yuri copied the saved waveform to a USB memory stick. She returned to her desk, inserted the USB memory into her computer, and opened the waveform file. She put the two waveforms side by side and looked at them on the screen. As she saw earlier, the waveform of the one that does not display the ERROR message has a slight wave at the falling edge of the signal.

"How was it?" Sugiyama said, looking at the computer display.

"The situation became a little clearer to me," Yuri explained the waveform.

"But I can't figure out why the waveform is wavy."

"Hmmm... ......" Sugiyama was also pondering.

"I don't get it. .... We need more information. But I wonder what we should look up."

"What's wrong?" Suzuki spoke up. Sugiyama explained the situation.

"Perhaps, it may be the ground," Suzuki said.

"Yes?" Yuri couldn't help but ask back at Suzuki's words.

"Let's measure the ground."

"Measure the ground? What do you mean?"

"Well, let's connect the probe's ground to the CPU's ground and measure the waveform of the ground of the radio IC. Although I don't know if we can see it."

"But the ground is zero volts."

"Yes, ideally."

Hearing this, Sugiyama said, "Oh, the ground bounce?"

"Yeah, maybe."

Yuri, too, finally understood what it meant.

"Ground bounce...... is a phenomenon in which the potential of the ground differs from place to place, right?"

"Yes, yes."

Yuri measured as Suzuki instructed. Again, she had to wait 10 minutes. When the trigger was applied 10 minutes later, a ground bounce had indeed occurred. There was a potential difference in the ground for a short period.

"But isn't the potential difference small compared to the rippling signal?" Yuri asked, looking at the waveform.

"There's bonding wire, too."

"Oh, yes."

The bonding wire is the wiring inside an IC and is a thin conductor that connects the terminals outside the IC package to the semiconductor chip inside the IC package. Even if there is a potential difference in the bonding wire, it is not included in this measurement waveform.

"But if the ground bounce is happening, the output change occurs on several signals simultaneously. ......" As Yuri said, the chime for the end of the workday rang.

"That's all for today." Suzuki's words brought today's survey to a close.

When the output terminal of one IC is connected to the input terminal of another IC, the ground is the reference for the signal's voltage level. Each of the two ICs has a ground terminal connected to the ground pattern on the printed circuit board. In other words, the ground pins of the two ICs are shorted together. Therefore, the voltage at the ground pins of both ICs are reference voltages, i.e., zero volts.

But that is the ideal case, i.e., when the impedance of the ground of the printed circuit board is zero. In reality, the ground impedance is not zero. Since there is some inductance, a potential difference proportional to the current change's speed is generated when the current changes.

When the output terminal voltage changes from high to low, a current flows from the output terminal to the ground inside that IC. The current from the ground terminal flows to the power supply's ground — usually the negative side. The inductance component of the ground generates a potential difference according to the time variation of the current — the amount of current change per unit time. If the potential of the ground terminal of the input IC is the same as that of the power supply side, the ground terminal of the output IC will have a higher voltage than the ground terminal of the input IC.

If the voltage on the output side — the potential difference between the output terminal and the ground terminal — is 1.00 volts, and the ground terminal of the output IC has a voltage 0.01 volts higher than the ground terminal of the input IC, then the voltage on the input side — the potential difference between the input terminal and the ground terminal — is 1.01 volts. This example's potential difference of 0.01 volts is due to the current's time variation. When the current becomes constant or zero, this potential difference will disappear. In other words, this potential difference is a pulse.

Thus, when the output signal falls, the ground current changes abruptly, and the ground inductance causes a potential difference, resulting in a pulse-like noise on the signal voltage at the input terminal.

However, a single signal drop is rarely a ground bounce problem. If it were, the signal would always be full of noise. Ground bounce actually becomes a problem when many output signals fall at the same time. In such cases, the larger the number of simultaneously falling signals, the larger the ground current, and the effect of ground bounce becomes non-negligible.

"Simultaneous changes in multiple signals ......"
That night, Yuri was in her room, thinking.

In this case, the wireless module Yuri was responsible for was the output side of the interrupt signal. She thought about simultaneous changes in multiple output signals, but the only signals other than interrupt signals were serial communication signals with the CPU. The serial communication signal did not work at this point.

"What are the other signals, if any .... Ah!" She couldn't help but shout.

She remembered the signals that might be the cause. She had forgotten about it because the NT500 and NR500 she was developing now did not use those signals.

"Oh, I did it. ......"
Yuri's mind went blank. But she had to figure out a way to verify it. Besides, until it is verified, it is not a definite cause. Frankly, she wanted to hang on to the possibility of finding another cause.

How shall I verify this .... Shall I modify the hardware or change the firmware ......

That night, Yuri couldn't sleep well.

I still wonder if that is the reason. .... It could be. I hope not. ......

The next day, Yuri modified the firmware of the wireless module first thing in the morning. A CPU controlled the entire repeater unit, and Numajiri handled its firmware. However, the radio IC used in the wireless module also had a built-in CPU. For simple control, the built-in CPU was sufficient. In fact, the peripheral unit was designed in this way. The peripheral unit's CPU built into the wireless module controlled everything; no external CPU existed.

But this was different with the repeater unit. This was because the computing power of the CPU in the radio IC was insufficient to manage the protocols for communicating with many peripheral units, especially the timing. Since the repeater unit used an external CPU, there was no need for a CPU built into the radio module. Using a radio IC without an internal CPU for the radio module was natural.

However, developing new radio circuits for each peripheral and repeater unit would take time and effort. Therefore, the repeater unit's radio module was made with the same circuit as the peripheral unit to shorten the schedule.

As a result, it was decided that the firmware of the CPU built into the repeater unit's wireless module would only perform very simple processing, and Yuri was in charge of that. Numajiri's load was heavy, and this was to distribute it.

The firmware that Yuri created did not process much. It operates according to commands sent by serial signals from the external CPU. There were three types of commands: transmit, receive, and setting.

In the transmit command, the given data was sent wirelessly. If a receive command came in, the radio module was put into the receiving state, and if a packet was received, an interrupt signal was output to the external CPU. When a request for received data came from the external CPU that accepted the interrupt, the received data was sent as serial data. In the setting commands, the communication frequency and other parameters were changed. It could have been better described as BIOS rather than firmware.

The radio module in the repeater unit had plenty of hardware to spare, as did the internal CPU's processing. Even after allocating IC pins for interrupt signals and serial data communication, there were still extra pins. Moreover, the external CPU also had extra terminals. Therefore, Yuri decided to implement a hidden function. Regarding hardware, she connected eight output ports of the wireless module and eight input ports of the external CPU. As for the firmware, she determined whether the incoming packet was an ERROR or ALIVE message and output the count value to the eight signals.

Six were used to count ERROR messages, and the other two were used to count ALIVE messages. As usual, the ERROR message count was made to count up 64 states from zero to sixty-three.

The ALIVE message counter had only two bits. So it counted from zero, and when it reached three, it did not count up any further ALIVE messages and held three. If an ERROR message came in that state, it was cleared to zero.

This was so the situation could be analyzed when communication breaks down in actual operation. However, the firmware on the external CPU side could not include a routine to process this signal due to the schedule, so it was not actually used.

In other words, the wireless module output the signals, but the external CPU did not see them. Nevertheless, if the firmware of the external CPU were to be changed on some occasions, the ability to send the status of these signals to the PC could have been added. Just preparing for that possibility was worth it, she had thought.

Yuri had implemented this function to analyze communication errors, such as the one that occurred this time, yet she had forgotten all about it. But it's still not a bad thing she forgot. It can be merely laughable. The problem was that this in itself seemed to be the cause of this problem.

When the sixty-third ERROR message came in, the ERROR counter was set to 63, or "all six bits are one." If three more ALIVE messages came in, the ALIVE counter was set to three, or "two bits are both one." If both happened, "all eight bits are one." If an ERROR message came in this state, the ERROR and ALIVE counter, i.e., all eight bits, were cleared.

All eight signals would change from one to zero or high to low. Moreover, an interrupt signal was also output at this time.

Everything made sense, considering that the eight counter signals would change from high to low when the interrupt signal fell and that this was the cause of the ground bounce.

The circumstantial evidence was perfect. There was no doubt that Yuri's firmware was the culprit. But proof, or verification, was needed.

Yuri had changed the firmware not to output the counter signals. She did the same measurement as the day before, and sure enough, the ERROR message did not drop out, and the ground bounce no longer occurred.

Now, the cause was confirmed, i.e., she was guilty.

I knew it was me. ......

Yuri sighed as reality hit her.

Oh dear. ......

She wanted to escape, but there was no escape. Closing her eyes would not change reality. Yuri took another deep breath and accepted the fact.

The solution should be firmware modification.

Yuri explained the cause and solution to Sugiyama and Suzuki. She was prepared to be scolded, but the response was surprisingly simple.

"Oh, so that's what this is about. I've never had a problem like this before."

"I see. I'm glad firmware modification can be the solution. It would have been a lot of work if it had been a circuit modification."

Sugiyama and Suzuki were satisfied, so Yuri emailed the cause and solution to all parties involved.

There was relief that the problem was solved, but the guilt she had caused was much more significant.

After a while, Yuri received an email from Shijo of Development Section 2. Shijo was in charge of hardware other than the repeater unit's radio module.

Sorry!
This interrupt input is not a Schmitt trigger.
I'm sorry!

That was right. This wouldn't have happened if it had been a Schmitt trigger input, but it couldn't be helped. It was that kind of CPU.

A Schmitt trigger is a digital signal input terminal with a hysteresis threshold, the boundary between high and low signals. In other words, the threshold has two levels: the lower one when the input is high and the higher one when the input is low.

This way, when the signal falls, the moment the high input becomes low, the moment the input voltage drops below the lower threshold, the threshold is switched to the higher threshold. As a result, the threshold is not exceeded even if the input voltage subsequently rises slightly. This eliminates the effect of pulsed noise, as seen in this case. Unfortunately, the interrupt input pin of the CPU used in this case was not a Schmitt trigger type.

During lunch break, Yuri looked for Ayaka with her tray in the company cafeteria. She looked at her usual spot and saw Ayaka waving at her. When Yuri sat beside Ayaka, Ayaka smiled and said, "You did it. Problem solved!"

"I did it, the cause of the problem," Yuri replied.

"You solved the problem. It's OK. You can say, 'I did it!'"

"Well, yeah. ......"

"Cheer up."

Ayaka started eating A-lunch, and Yuri began to eat pasta. Yuri remembered. When Ayaka emailed her a picture of Kirishima's business card, she said, "I did it." Yuri wondered how Ayaka had been doing with Kirishima since then. Yuri was curious but did not feel like asking Ayaka about it.

Ayaka, who had finished eating her A-lunch, said as she sipped her tea. "This glitch was cleared up quickly, considering we went through the whole system."

"Yeah, I guess so."

"You don't look well."

Yuri was even less energetic because she remembered Kirishima, but Ayaka did not know that.

"Oh yeah, let's go out for a drink today. It's been a while. I don't have any plans today, I think. ......" Ayaka said, pulling out a notebook from her pocket and checking her schedule.

Then, two pieces of paper fell from the notebook to the floor.

"Something fell out," Yuri said and picked up them. Yuri's eyes were glued to them. They were Kirishima's business card, torn in half.

"Ah, ..." Ayaka raised her voice. She looked uncomfortable.

"What is this? What does it mean?" Yuri couldn't help but ask.

"Well, Let's see. ......" Ayaka was depressed.

At that moment, the chime rang. The lunch break would be over in five minutes. They had to return to work.

Yuri stood up from her chair and said.

"Well, let me hear it later over a drink."

"OK."

After work that day, Yuri and Ayaka met at the company gate. Yuri was waiting for her, and Ayaka soon arrived as well. Yuri raised her hand to signal, and Ayaka silently raised her hand. They got on the bus together and remained silent. They got off the bus before the station and entered an izakaya, a tavern.

As they took their seats and ordered beers and snacks, Yuri started a conversation.
"So, what's it all about?"
"Yeah, um, ..."
Ayaka began to talk about the day she emailed Yuri the picture of Kirishima's business card.

Ayaka was at an exhibition that day. Although she pre-registered every year, she was often unable to go. This year, however, she got some time to go. The exhibit included wireless LAN, RF-ID, logistics and inventory management systems, and various devices, modules, and components. She had seen the whole exhibition and had some time before its end, but she decided to leave before the train got crowded on her way home.

At that moment, Ayaka's eyes caught sight of Kirishima. The profile looking at the exhibit was definitely Kirishima. He was as cool as ever. Ayaka reflexively shouted, "Kirishima-san!"
Kirishima turned toward her. Ayaka ran toward Kirishima. Kirishima looked dubious and tilted his head.
Oh, shit.
Ayaka recalled that she had never spoken to Kirishima one-on-one before. She had only greeted him when she was with Yuri.

What should I do?

But Kirishima was watching her steadily. Ayaka was also looking at Kirishima, smiling and waving as she ran. She could not now pretend that nothing had happened. Kirishima was getting closer to Ayaka, almost there. If this happened, it was only going to happen. Ayaka stopped in front of Kirishima, determined.

"Kirishima-san, it's been a while," Ayaka greeted him as if he knew her.

"Uh, ..." Naturally, Kirishima was puzzled.

He couldn't say he didn't know her because it would be bad if Kirishima had forgotten someone he knew. So, there was an opportunity for Ayaka.

"Oh, I'm sorry, you don't remember me. I am Ayaka Nanase. I am Yuri Ueda's peer."

Ayaka answered with dignity. She was not lying about anything.

"Oh, you are Ueda-san's peer ...... Nanase-san."

Kirishima seemed a little relieved.

Yuri's name worked. Now, Ayaka won't be seen as a suspicious one.

At the exhibition, they wear an admission card with their names printed around their necks. Therefore, just because someone calls your name, you cannot tell if you know that person.

"You look well. Are you now working in mobile telecommunications infrastructure?"

"Um, ... sorry, I don't know if I could talk to you how much."

"Oh, that's right. With development work, you can't talk." Ayaka replied.

This is troubling. What do I do if I can't talk about the job?

She had to escape to a common topic.

"Yuri is fine, too."

"Really? How is she?"

Ayaka was made to realize that Kirishima was interested in Yuri's.

"I'm a software engineer, so I don't know much about it, but the products Yuri was in charge of have been mass-produced and shipped."

"The one she designed last year?"

"Well, it's a slightly modified version, a product for another project."

"Hmmm. Well, Ueda-san really does work hard."

Ah, so all he talks about is Yuri.

"I'm sorry, but I just got here and haven't seen most of it yet. I want to hurry up and look around," Kirishima said hurriedly.

"I understand. I'm sorry to bother you."

"No, no, it was great to hear how Ueda-san is doing. ...... Can you give this to Ueda-san?"

Kirishima then offered his business card to Ayaka. She took it and offered her own business card.

"OK. And here is mine."

This would be considered good manners, Ayaka thought, rather than an imposition.

"Thanks."

Kirishima accepted her business card, but it was obvious that he thought it would be rude not to.

"Bye," Kirishima raised his hand and started to walk away.

"Goodbye," Ayaka bowed and sent Kirishima off.

Kirishima disappeared into the crowd.

I'll email Yuri a picture of his business card.

Ayaka took a picture of Kirishima's business card, emailed it to Yuri, and left the venue.

Walking toward the station, Ayaka looked at Kirishima's business card. She had only said hello to him once. She was embarrassed at herself for approaching him so casually. Ayaka did not particularly love Kirishima. She had no desire to date him. She just wanted to have a friendly talk with him. But since they had never spoken, it was only natural that Kirishima's reaction to her would be muted. And all he talked about was Yuri.

Ayaka had never felt this miserable before. She had usually taken the time to tell men she wanted to get to know that she was interested in them in a casual way and let them approach her. Ayaka had never failed at it before. This time, she didn't even confess — and before that, she didn't even want to date him — but she felt like she had been rejected for the first time.

Ayaka knew she was the cause, so she felt miserable and angry with herself.

Damn it! Ayaka involuntarily tore up the business card.

Oh, no! Oh shit! I had to give this to Yuri. ...... Ayaka felt lousy.

I've already emailed her the picture, so I'll have to settle for that. Ayaka had to console herself.

"Upon this, I have no choice but to make Yuri happy. She denied it, but there was no doubt that she had feelings for Kirishima. He was also concerned about Yuri. OK, I'll call Yuri," Ayaka thought.

But before she could call, she noticed an email had arrived. It had been forwarded to her from her company email address. The email was about a communication error. It was from Shinomiya Foods, saying that the error was regular.

Wow, does this mean my app has a bug? Ayaka had to call Yuri even more.

"...... And, well, that's what it was all about," Ayaka sighed as she finished her explanation.

Yuri was speechless, but Ayaka was looking upward at Yuri, and Yuri searched for the right words.

"Words fail me," It was the word that came out.

Yuri did not know if those words were appropriate for Ayaka, but she meant them.

"You're right," Ayaka admitted and downed her glass of beer.

Another sigh, and now Ayaka asked a question.

"So, how about you?"

"How's what?"

"Did you send an email?"

"What are you talking about?"

"I said, did you email Kirishima-san?"

"Of course not."

"What? Why?"

"Because I thought ...... Ayaka was gonna go out with Kirishima-san."

"What are you talking about? I'm not thinking about that at all."

"But you said he is cool."

"But that doesn't mean ......"

"You never give me his card, just a picture."

"That's true, but ......"

"And you said, 'I did it.'"

"What? It means I sent Yuri the email and picture, so now it's your turn."

"I thought it meant that you got the card, what you wanted."

They both sighed. That was funny and made them laugh. It made her feel better, and Yuri asked Ayaka what was on her mind.

"But why did you email me a picture? Normally, email addresses are exchanged via infrared."

Yuri's cell phone could exchange contact information via infrared communication. Japanese cell phones of the time had that function.

"Because Kirishima-san gave me his card, And mine is ...... this," Ayaka pulled an iPhone out of her bag.

"You got an iPhone."

"Yeah."

"Let me see."

"Don't drop it. It doesn't have a strap."

Ayaka turned on her iPhone and then handed it to Yuri.

"Why don't you just put it in a case?"

"It's not cool."

Yuri manipulated the screen and looked at the icons.

"You've got a lot of apps. ......, you've got ebooks, too."

"Yeah."

When Yuri returned the iPhone, Ayaka pressed the sleep button and put it back in her bag.

"Hey, Yuri, why don't you email Kirishima-san?"

"I won't. ......"

"But it's hard for Kirishima-san to email you."

"Why?"

"It's natural. Because Kirishima-san left the company, Yuri was transferred to Section 1."

"Eh?"

"Hm? You don't know?"

"I don't know. ......"

"Words fail me."

Yuri recalled Kirishima's words, "But after teaching you radiofrequency, I realized ......"

This might mean that Kirishima's leaving had been decided before Yuri moved. If so, Yuri could understand what Ayaka was saying.

"I didn't know that," Yuri sighed.

"Maybe you don't know about the promotion either?"

"What?"

"It's the Tokoroten Method."

"Huh?"

"Just as I thought. You don't know."

"What is the Tokoroten Method?"

"When one person goes in under, another person goes up."

"Hmm?"

"You know we had a new guy in the Systems Development Section last year, right?"

"Yeah."

"So, in this year's evaluation, he got a C, and I got an A."

"What?"

"Yuri, you also got a C in your first year, didn't you?"

"Eh ......"

That was a story she never told anyone.

"I got a C my first year, too."

"Really?"

"In a nutshell, it's a performance-based system in form, but it's operated based on seniority."

"Oh, shit."

"Yuri's got a new hire this year, so you will be promoted next year."

"I feel like they're idiots."

"Well. I heard that they used to operate on a performance-based system, but there were a lot of negative effects, so they just cheated on the operation for the time being. And that seems to be the way it continues."

"Hmmm ......? Hey, did you know that from last year?"

"Last year, ..., I vaguely knew about it."

"So, that's why," Yuri nodded.

"What?"

"You didn't seem to be interested in your evaluation."

"Oh, yeah."

Yuri began to feel like the one who was being an idiot.

The solution

Yuri received an email from Inada, a sales representative. It was about how to proceed with the solution to the problem.

The solution was to rewrite the firmware of the radio module in the repeater unit. Still, instead of rewriting the installed repeater unit, the procedure was to rewrite another, bring it to the site, and replace it.

Since the company did not have enough repeater units, spare units at the customer's site would be sent to the company to be rewritten. Once the spare unit arrived, the rewrite could have been done immediately, but the on-site replacement would be done the following week. It was not impossible to work while the line was running, but it was decided to work on a day when the line was stopped. It was easier to work that way.

First, they had to rewrite the firmware of the repeater unit in the company, and then they had to wait for the spare unit to arrive.

"Or should I rewrite all the firmware at once after the spare unit arrives?" Yuri was pondering when Kishimoto spoke to her.

"Ueda-san, may I have a moment?"

"Uh, yeah."

"Smith Chart, I've been thinking about it in my own way," Kishimoto said, holding the book and opening the pages.

Oh shoot, I totally forgot about it. I have left him alone, Yuri thought.

Still, Yuri remained calm and answered.

"How was it?"

"I think I understand. This equation represents the relationship between the reflection coefficient and impedance, right? So I think the essence of the Smith chart lies not in the impedance, but in the reflection coefficient."

"Hm. ...."

Yuri thought that Kishimoto seemed to have found the essence of the matter on his own.

"So?"

"So ...... maybe ...... the points on the Smith chart represent points on the complex plane of reflection coefficients, I'm thinking."

Yuri nodded silently and waited for the rest of the story.

"Well, ... and I think the Smith chart shows the impedance of that point, the point representing the reflection coefficient."

"Briefly?"

"Briefly ...... I think the Smith chart is a complex plane of reflection coefficients with a scale of impedance on it."

Amazing. I'd say it's the correct answer. But hopefully, a little more, Yuri thought.

"And so?"

"Eh? Oh, why use Smith charts, right? Why? First, it is to take impedance matching. And that is to reduce the reflection coefficient. So, it is to make the reflection coefficient a diagram and to make the relationship between it and impedance visually clear."

"That's great," Yuri said. "It's perfect."

Yuri couldn't help but applaud.

"Really?" Kishimoto said.

"Yes. Really."

"Oh, ..., I'm so happy."

Kishimoto scratched his head, perhaps embarrassed.

"I couldn't figure it out that well alone," Yuri added.

"Really? But it was thanks to you, Ueda-san."

"Why?"

"Because you asked me not to use formulas, or why we use that."

"Yeah."

"It's because of that tip. I never thought of it that way."

"Oh, well, ..., that's good."

  She had no idea that he would be so pleased. Yuri was much happier that Kishimoto was delighted than he could understand the Smith chart.

"Kishimoto-san, you are amazing!" Suzuki spoke to him.

"Thank you."

"Ueda-san, you're great too. You are indeed very good at teaching."

"No, I'm not."

"Yes, you are. Because they say, 'A person will not move unless you show them to do, tell them, have them do it, and then praise them,' and you did it."

"What's that?"

"Isoroku Yamamoto, isn't it?" Sugiyama answered Yuri's question.

Isoroku Yamamoto? I've heard that name before. ......

"Yes. You did it," Suzuki nodded.

"No, I haven't been able to do it since I 'show them to do' in the first place."

"You did it to 'show' why it's a sine wave."

"That's......." *I just wanted to figure out my question.*

"Then 'told' him to think for himself why ......"

It is my Dad's word.

"And you 'had him think by himself'......"

It is also my Dad's word.

"And you even applauded and 'praised him.' It is wonderful."

I just really thought it was amazing. After all, I couldn't have done it.

But she decided to be honest here and take his word for it.

"Thank you."

'A person will not move unless you show them to do, tell them, have them do it, and then praise them.' it's a good word.

At that moment, Yuri realized. Suzuki showed Yuri, having Kirishima teach her. He told her not to over-teach. He had Yuri teach Kishimoto. And then he praised her.

After all, I am in the palm of Buddha's—or rather, Suzuki-san's—hand.

Yuri realized once again how inexperienced she was. At the same time, however, she thought that even a person like her could do this job.

The following week, the repeater units were replaced at the Shinomiya Foods plant. They were mounted in high places, mostly on cable ladders.

Yuri, who had caused the defect, wanted to do the work herself. Still, Inada, the sales representative, and Shibata, the manager of Development Section 2 and the project leader, did the work because they could not allow a woman to do it. Yuri was a little frustrated, but she complied.

Instead, Yuri was prepared to take on the role of being scolded by the customer. Koyama of Shinomiya Foods' engineering department was present at the work. Yuri apologized to Koyama as she watched Inada and Shibata work.

"I apologize for any inconvenience this may have caused you."

Yuri expected Koyama to yell at her or at least complain. However, Koyama uttered words that surprised her.

"No, I'm not inconvenienced at all. I'm sorry for the mean-spirited test."

"Well, What do you mean?"

"In actual operation, we haven't had any communication errors. This is because printer errors do not occur that often. So, we intentionally prepared a print overfill, mixed it up on the line, and used it again when it went through the image recognition equipment so that the error message would appear repeatedly, and then we tested it."

"But that's much work."

"Well, that's part of the engineering department's job. Besides, there's a regularity to it, which is a relief."

Regularity means that somewhere in the system, there is a cause of the malfunction. Surprisingly, despite the glitches, he was relieved instead of complaining.

"Relief?" Yuri asked.

"Yes. If we know that the error message is not received once in sixty-four times, then when the sixty-third time comes up, we can intentionally cause the sixty-fourth error by putting in the filled-in one that I mentioned earlier, and we will be OK for the time being."

"I see! So if you intentionally make errors, errors won't be a problem."

Yuri had never thought of that. They were professionals in the field and could make it work under the given conditions.

"Yes, that's right. And this system has helped us a lot."

"What do you mean by that?"

"Now, we are streamlining and don't have as many people. Of course, we have enough people to deal with any mechanical breakdowns or other problems. But the problem is that we learn about the glitch immediately."

"But bad printing is repelled, right?"

"Yes, but if the printing defects continue, many defective products will remain. Even if there is a problem here, products flow from the upstream process without concern. You can't stop in the middle once you start making a food line."

"Oh, I see. ...... It's not the same as manufacturing a machine."

"That's right. So we would need people to confirm that there are no glitches, even though we would need fewer people if there were no glitches."

"It's kind of ironic, isn't it?"

"You're right. But this system minimizes the number of people checking, which is a big help."

Yuri was hesitant to hear this but asked what she was wondering.

"Are you going further to reduce the number of people with this system?"

Koyama replied with a laugh.

"No, no, no. If we reduce the number of people further, the cost savings will be negligible. Instead, we are trying to use the spare human resources for other things. Some people regularly participate in product planning at the head office or have someone from the head office come in and explain the situation on the ground. That has made the communication with the head office much more open. Previously, people in the field would say, 'Headquarters doesn't understand the field,' and headquarters would say, 'The field doesn't listen to us,' but that has almost disappeared."

"I understand. I'm glad we could be of help."

Rationalization by machines or labor-saving should not be denied. Rather, it is often the reason domestic factories can survive. Nevertheless, seeing people reduced by one's own products is unpleasant, especially in the current economic situation. Yuri was relieved that this system would not be used to reduce the number of workers.

After about two hours, the repeater unit was successfully replaced, and a test run confirmed that the problem had been resolved.

The following day, as Yuri got off the bus and walked to work, she noticed Ayaka was a few people ahead of her. Ayaka was listening to some music on her earphones.

"Good morning."

Yuri tapped Ayaka on the shoulder, and Ayaka removed her earphones.

"Oh, good morning."

"What are you listening to?"

"Well, ... 'Moonlight'"

"Huh? Ayaka listen to Beethoven?"

When Yuri said that much, she noticed Ayaka grinning faintly.

"Not, apparently," Yuri added.

Hearing this, Ayaka burst out laughing.

"Ha-ha-ha, you guessed it. It is Chihiro Onitsuka."

"I see."

Ayaka turned a straight face and asked Yuri.

"Did you email Kirishima-san?"

"No."

"You should."

"But ...."

"Think about it. This is the best situation."

What does that mean?

Before Yuri could voice her doubts, Ayaka continued.

"If you were dumped, you wouldn't have to see him. But if it were someone in our company, you wouldn't be able to do so. If you were in the same office, you would see each other daily, even if you were dumped. With Yuri, that's not the case."

Ayaka thought that way. She is right.

"But what would I write?"

"You just let him know how you're doing. You don't have to write him 'I miss you.' The important thing is to give Yuri's mobile phone email address to Kirishima-san."

I see. ......

The two went straight to their respective workplaces.

As usual, Yuri was with Ayaka for lunch, but they did not talk much. Yuri was hesitant, and perhaps Ayaka had said what she needed to say, so she didn't want to mention the email.

But as soon as they finished eating, Ayaka, next to Yuri, cut to the chase.

"So, did you email him?"

"No."

"You're so laborious."

Saying this, Ayaka suddenly took the cell phone out of Yuri's pocket.

"Ah!"

When Yuri noticed, Ayaka was already operating Yuri's cell phone.

"I know your cell phone. Oh, what's this? There is something saved."

"Stop."

Yuri tried to take the phone back, but Ayaka dodged her, and Yuri's hand cut through the air.

"You drafted it. ...... Yeah, that's fine. It's all right. All you have to do is send it."

"Give it back."

"Shall I send it for you?"

"Stop."

"OK, OK."

Ayaka returned the cell phone to Yuri.

"Send it."

"But ......"

"Kirishima-san has been waiting for you for days now."

"What?"

"Kirishima-san gave me his business card for you, so you're supposed to contact him."

"......"

"Do it, or you'll regret it."

"……"

"Now!"

Yuri took the plunge and sent the email.

"Good! Well done!"

Ayaka patted Yuri on the back.

"Well, now we'll just have to wait and see when we get a response: …… in a few minutes, in the evening, or tomorrow: ……"

"You look fun."

"Yeah, because it's none of my business. Hey, Let's have a cup of coffee and wait for his reply."

They moved to the coffee corner, and each grabbed a cup of coffee and sat at a table.

"I wonder what the reply will be," Ayaka said, holding her coffee cup and looking up to the heavens.

"You look really happy," Yuri said.

"Yes. I am."

At that moment, Yuri noticed that Ayaka wore a ring. It was the ring finger of her left hand.

"Hey, that ring ……"

"Oh, you finally noticed?"

Ayaka unfolded her left hand to show the ring to Yuri.

"What? What do you mean?"

"That's what I mean."

"With who?"

"With him."

"Why?"

"Because he proposed to me."

"What? I thought you were breaking up with him."

"Well, on our other weekend date, I said, 'What do you think about me?' And then ......"

"And then?"

"He said, 'This is what I think!' and handed me this."

"Oh."

"There's no mood, nothing at all. He had been wondering when to say it out."

"Hmmm ...... silly."

"I agree. Haha ......"

At that moment, a chime rang. The lunch break would be over in five minutes.

"I knew the reply wouldn't come so soon," Ayaka said.

They left their seats, returned their coffee cups, and headed for the elevator to return to work. Then Yuri's cell phone rang a notification. Yuri rushed to take it out and open it.

"You got the reply!" Ayaka said, peering into the screen.

Yuri opened the email, making sure Ayaka could not see the screen. Ayaka looked dissatisfied but waited patiently.

"How was it?"

"......"

"Hey."

Yuri silently showed the screen to Ayaka.

Ayaka read the email in a great hurry.

"Good for you!" Ayaka hugged Yuri.

Yuri remained silent and nodded.

I guess I haven't heard from Kirishima-san until now because I haven't contacted him. Come to think of it, there were a lot of possible causes for the wireless glitch, but surprisingly, it was me.

Yuri's smile turned into a wry smile. But it was not unpleasant. She was in the mood to have learned a good lesson.

Yuri had one word left to say.

"Best wishes, Ayaka."

"Thank you."

Glossary

design review
In product development, for example, the steps are planning, design, prototyping, mass production prototyping, and mass production, and design reviews are held at the end of each of these steps. The design review is attended by the person in charge of the product from each section, such as development, sales, quality assurance, and the technical department of the mass production plant. The project's status is reported, problems are pointed out and discussed, and a decision is made as to whether or not the project can proceed to the next step.

DIP switch, two poles, four poles
DIP switches are multiple switches combined into a single component. They are called DIP switches because they have the same shape and dimensions as DIP (Dual In-line Package) IC packages. The number of switches in a DIP switch is called the number of poles. Two switches are counted as two poles, four switches as four poles, and so on.

CPU
CPU stands for Central Processing Unit. The CPU is the heart of a computer. This part executes programs. In old mainframe computers, the CPU was about the size of a chest of drawers. Memory and input and output devices are connected to the CPU.
A microprocessor is a CPU made into a single IC (Integrated Circuit, semiconductor component). A microprocessor is also sometimes called a CPU. Some microprocessors contain a CPU, memory, and input and output circuits in a single semiconductor component. In this novel, such

microprocessors are called CPUs.

I/O port
Input/output digital signal pins of a microprocessor equipped with input/output circuits. Through these pins, signals are input from or output to the outside of that IC.

AD converter
A circuit that converts an analog signal into a digital signal (numerical value).

DA converter
A circuit that converts digital (numeric) signals into analog signals.

Buffer IC
This IC outputs the same signal (voltage) as the input, and by using this IC, the circuit on the output side can avoid influencing the circuit on the input side.

Ground
The reference potential of an entire circuit or its wiring.

Threshold
A digital signal with a voltage lower than the threshold is identified as low level (0), and a digital signal with a voltage higher than the threshold is identified as high level (1).

Noise Margin
When noise is applied to a digital signal, if the signal's voltage does not reach

the threshold, the value (0 or 1) as a digital signal remains unchanged. This margin, or the potential difference between the signal's voltage in the absence of noise and the threshold voltage, is called the noise margin.

Pattern

Electronic device circuits are usually fabricated as printed circuit boards. A printed circuit board is an insulator board with a conductor wiring pattern formed on it. Electronic circuits are completed by soldering electronic components to the board. The "pattern" is the conductor wiring pattern already created on the printed circuit board.

Capacitor

An electrical component that stores an electric charge. It does not allow DC to pass through, but AC tends to flow more easily at higher frequencies.

Decibel

A notation used by electronics engineers for the ratio of two powers. The power of one is divided by the base power of the other and then multiplied by ten after taking the ordinary logarithm. A ratio of 10 times is plus 10 decibels and a ratio of 100 times is plus 20 decibels.

Matching

To make the impedance of the two connected circuits the same. In radio-frequency circuits, matching is important because signal power cannot be effectively transferred unless the impedances match.

Inductor

An electrical component made by winding a conductor. It allows DC to pass through, while AC tends to flow more easily at lower frequencies.

Reflection coefficient
When radio-frequency signals are input from outside the circuit, part of the signal is generally reflected back. The reflection coefficient is the ratio of the reflected signal voltage to the input signal voltage. Since the signal is AC, it is a quantity consisting of the ratio of amplitude and phase difference. The impedance of the input and output circuits determines the reflection coefficient.

Transistor
An amplifying element made of semiconductors and is also used as a switch.

Reverse phase / Inverse phase
The phase which is opposite to some AC signal, i.e., the phase difference is 180 degrees.

Emitter, Collector, Base
The name of the transistor terminals, respectively. An emitter-grounded amplifier circuit is an amplifier circuit in which the emitter of the transistor is the reference potential, the base is the input, and the collector is the output.

Intermediate frequency
A typical receiver converts the received frequency to a lower frequency before the signal is processed (demodulated). This converted frequency is called the intermediate frequency.

Through-Hole

A hole drilled in a printed circuit board, the inner wall of which is a conductor pattern. It electrically connects the wiring patterns on the front and back sides.

Return loss

The ratio of reflected signal power to input signal power when a portion of the signal input to a circuit is reflected. Usually expressed in decibels.

Farad, Henry

Farad is a unit for quantitatively expressing the properties of a capacitor (capacitance), and Henry is a unit for quantitatively expressing the properties of an inductor (inductance). They correspond to ohm in the case of resistance.

Overshoot, Undershoot

When the voltage level of a digital signal goes from low to high, the voltage may momentarily exceed the steady level of high and settle to a steady level of high over time. This momentary exceeding of the steady-state level of high is called overshoot. Conversely, when going from high to low, a voltage momentarily lower than the steady-state level of low is called undershoot. These may cause malfunction or failure of digital circuits.

Preamble

An additional signal added before the main body of data when transmitting digital signals by radio. Typically, a repeating signal of 1010...... is used. The receiver receives this signal, stabilizing the circuit state and preparing it to receive the main body of data. In a digital signal receiver, when no received signal is present, the input is only noise, and 1s and 0s are randomly output at the receiver output (demodulation output). Still, detecting the preamble's signal pattern makes it possible to know that a received signal is present.

EEPROM

EEPROM stands for Electrically Erasable PROM; PROM stands for Programmable Read Only Memory. Electronic devices can write data to EEPROM during operation, which is retained even when power is turned off.

SAW Filter

Surface Acoustic Wave filter. A filter that uses vibrations on the surface of a piezoelectric material to achieve higher performance than a filter using inductors and capacitors.

Trap

A filter that blocks the passage of only a specific frequency. A series resonant circuit of an inductor and a capacitor is placed between the signal and the ground.

Prefix (Auxiliary Unit) example

Mega: 10^6, or one million
Kilo: 10^3, or one thousand
Deci: 10^{-1}, or 1/10
Nano: 10^{-9}, or 1 / 1 billion
Pico: 10^{-12}, or 1 / 1 trillion

CSV

Comma-Separated Values. A CSV format file is a text file with values (data) separated by commas and the extension CSV. It can be opened in an editor or Excel.

Interrupt

To execute a particular process (interrupt routine) during the execution of a program due to some factor. Examples of factors that can generate interrupts include signals input from outside the CPU and timing created by timers built into the CPU. When processing of the interrupt routine is completed, program execution returns to the process before the interrupt.

Afterword

This book contains two works: the first half, 'Impedance Matching,' and the second half, '50 Causes of Arriving No Messages.'
'Impedance Matching' was published by ONBOOK as an on-demand publication in October 2007, and '50 Causes of Arriving No Messages' is a sequel to that work, which is the first publication of this book.

When I published 'Impedance Matching,' I received words like "a novel that had never been done before" and "a work that is a bit different." These were very gratifying evaluations for me as the author. Because the reason I wrote this piece was that I wanted such a novel to appear, but it didn't. So, if they had said it was "a novel similar to something" or "a work in an existing genre" (even if they said it was interesting), it would have been a failure for me.

The "such a novel" here are novels in which the main character is an engineer and has realistic technology-related episodes. I had been craving such a novel. I coveted such a novel because I am an engineer myself. Many books, movies, and dramas are based on the main character's jobs, such as lawyers, detectives, doctors, salesmen, manga artists, etc. However, I know of very few works that depict the work of engineers. Even on the rare occasions when such works existed, they were primarily business rather than technical. That was unfortunate.

I am an electronics engineer working for a Japanese manufacturer. Since joining the company, I have been developing electronic circuits and software. Currently, I am mainly responsible for designing radio-frequency circuits for wireless devices.

I first became interested in electronic circuits in the third grade of elementary school when I came across 'Electronic Block.' I still clearly remember the images I saw when I first encountered them at Sanseido's main store in Ochanomizu, Tokyo. I saved my allowance and combined it with my New Year's gift to buy an 'Electronic Block.'

After that, I built radio kits. In the sixth grade of elementary school, I obtained a telephone class (today's fourth class) amateur radio operator's license. I opened my station in the second grade of middle school. In the third year of middle school, I tried to build a microwave transmitter, but I started high school before I could get it working.

In high school, I planned to join the amateur radio club. There was a club recruitment day on a Saturday afternoon right after I entered the school, and there I found a club called the 'Computer Research Club.' I was surprised because I hadn't expected a computer club. After all, this was a time when personal computers did not yet exist, and calculators were expensive. I did not hesitate to join the 'Computer Research Club.'

The high school was a regular school, not a technical or commercial high school, so there were no mainframe computers. However, there were a few old programmable calculators that were always available. Still, the main computers we used in this club were mainframe computers. We went to technical or commercial high school and were allowed to use IBM and FACOM (Fujitsu) to run FORTRAN. It could take more than an hour to travel one way and was available only for 30 minutes to use the mainframe. We borrowed a mini-computer from a computer manufacturer at the fall school festival and demonstrated the program.

Thus, I programmed for a while but gradually became interested in hardware. I was naturally self-taught since there were no seniors or teachers to design and build the hardware. During the summer vacation of my second year in high school, I designed and built a circuit to add and subtract single-digit decimal numbers, using only TTL and no LSI. In my third year, I built a microcomputer using a 4-bit CPU, which was included in the book 'Tsukuru Computer' (in English 'Making a Computer'). When I found this book in a bookstore, I was thrilled to see that the time had come for individuals to build their computers! A few months later, the famous Japanese CPU training kit, the TK-80, went on sale. Later, my addition/subtraction circuit and 4-bit microcomputer were introduced in several magazines, including 'CQ ham radio.'

When I entered college, I started working part-time in May at a company as a programmer. The language was COBOL. This part-time job was a great learning experience. I am now comfortable with extensive programs and have gained confidence that even if there is a bug, I can solve it by thinking logically. Incidentally, at that time, we were still using punch cards, and personal computers were starting to come into the world, such as Apple II, PET, TRS-80, etc. However, the general public has not widely used the term "personal computer" for a while.

I bought a TRS-80 with the money I saved from my part-time job. One friend had an Apple II, and another had a PET. I made and sold a game program for the TRS-80. I remember porting a LISP interpreter (from the 8080 to the Z80) that appeared in a magazine called 'Dr. Dobb's Journal', which was a fun and educational experience.

After starting work, I was assigned to a department that developed embedded devices. For the first year, I was in charge of circuits related to video signals, followed by circuits and programs (firmware) for controlling mechanisms.

At that time, the scale of firmware was small, so control circuit designers wrote the firmware. After a few years, the firmware grew so large that different people handled control circuits and firmware design. I was in charge of control circuit design and started designing gate arrays (custom LSI).

For over a decade, I designed control circuits for various models and had always worked with digital circuits. Still, about ten years ago, I suddenly became in charge of designing radio-frequency circuits for wireless equipment. I joined a project for a wireless-based system midway through the project.

The radio part had been designed outside the company, but it didn't work out, and they turned the job over to me, saying, "Since you used to do amateur radio, you can do it."

At that time, my understanding of radio frequencies was "I know the terminology. I know the block diagrams. I have seen schematics. But I can't design circuits." Inwardly, I wondered if I could design radio-frequency circuits with the same feeling as the main character in the novel.

I started by buying and reading books. I had seen the Smith Chart since elementary school, but when I read the book, I felt that "I understand the words written in the book, but it doesn't ring a bell." After three days of thinking, I finally understood, "Well, it's not a theory. It's a tool."

Later, a radio-frequency engineer from another company was assigned to design the circuit, and I studied with him for about a year. Thanks to this, I have managed to design subsequent products independently.

Next, I would like to talk about my relationship with fiction. When I was a student, I read only science fiction in fiction. These included works by Arthur C. Clarke and Shin'ichi Hoshi. When I was a student and first joined the company, I thought, "Someday, I would like to write a novel myself." I had science fiction in mind then but somehow forgot I wanted to write a novel. Becoming an electronics engineer was my childhood dream, and writing novels was a secondary dream.

After getting a job and becoming an electronics engineer, I was so busy with work that I forgot I even had a second dream. However, as I have watched movies and read novels over the past few years, I have come to think that very few works depict engineers. I wondered if someone would make such a work.

At the same time, as I began to work with radio-frequency circuits, I began to crave a book that explained them more clearly. I wanted a book written not from the perspective of the resulting theoretical system but from how our predecessors developed their ideas.

In the two pieces in this book, I explain the meaning of Smith charts, decibels, oscillating circuits, and sine waves, which I arrived at by continually asking myself, "What does this all mean?" While theoretical systems are essential, I think it is also precious to grasp physical phenomena intuitively.

In the case of a simple circuit, it may be possible to find a mathematical solution for how it behaves if you can formulate an equation. Even if the immediate goal has been achieved by finding a solution, it is difficult to apply the solution to other problems unless one can imagine "why such a solution is obtained."

As I thought I wanted such a "work depicting engineers" and a "book explaining high-frequency waves in an easy-to-understand manner," I began to develop a vague image of how I would write such a book. But I didn't intend to write a book about it myself; I thought, "Will someone please write it for me?"

I then came across a film, 'Switch to Happiness,' written and directed by Mana Yasuda. The film is set in a rural electronics shop, which intrigued me as an electronics engineer. I was then surprised to learn about Director Yasuda's background through information before the film's release.

Director Yasuda started making films when she was in college. She worked for an electronics manufacturer and produced at least one film yearly while working as an office worker for about ten years. She submitted her films to various film festivals and won awards. I was amazed that one could do such a thing while working as an office worker.

I thought, "That's one way to do it," and decided to write my book. Well, it wasn't so much that I "made up my mind." Rather, when I realized it, I was ready to do it. In Japanese, an expression says, "push someone's back." It means "encourage someone." I felt like I had been "jump kicked in the back." I remembered, "Come to think of it, I used to want to write a novel." By the way, the movie 'Switch to Happiness' is a tasteful and lovely movie, and I hope many people will watch it since it is now available on DVD.

After deciding to write my first novel, I started putting ideas together, mainly on my days off. What kind of story should be told, what technical content should be included, and how should the difficulties and joys of engineers be portrayed? It took me about six months to decide on the plot, two months to write the first draft, and another three months to polish it before sending it to the publisher.

Usually, if you wanted to publish a book, you would take your project or manuscript to a publisher, but I did not take that approach. Since the author is an unknown company employee and the work is an "engineer's novel," the publisher might say it is not for the general public. As the author, I desired to make the book enjoyable for the general public, but more than that, I prioritized making it enjoyable for technical people. That is precisely why I wrote this piece; it would be meaningless if the work were rewritten to make it more suitable for the general public and less enjoyable for technical people.

Therefore, I initially considered self-publishing. However, self-publishing costs about the same as buying a new car. So I searched the

Internet and found a publisher called ONBOOK. Since it was an on-demand publication, there were very few initial costs. However, I had to create the final data for the text and cover and convert it to PDF by myself. The first piece had lyrics, so I had to get permission to use them, which I did myself. It took a lot of time and effort, but it was a good experience.

After the first book was published, I gave a few copies as gifts to acquaintances and celebrities, and people I know online purchased them and commented favorably on them. I have heard that one company purchased one copy for all board members, and another purchased a batch for training new employees. Thank you very much.

Immediately after publishing the first book, I received requests for a sequel, but at that point, I did not know whether I would write a sequel. However, one technical episode remained unwritten. That is the meaning of the sine wave, written in the second piece. It was in the back of my mind for a long time, and about a year after the first book was published, I decided to write a sequel. The sequel also took about a year to write.

One of the things I thought a lot about during the two works was the technical episode. It had to be interesting or informative for the engineers to read yet not violate the company's confidentiality.

I also omitted the supplemental information that would have been needed if I had written it properly because too much technical precision would have made the novel redundant. Therefore, I know the reader may want to make a quip in some places.

Nevertheless, I have tried to balance being attractive enough to be read by engineers and having enough (if not perfect) technical reality.

For example, the circuit that appears at the beginning of the first work, which uses one AD converter to read two switches, has extreme resistance values. The work's ratio is "one kilo-ohm and two kilo-ohms" versus "two megohms and one megohm," a thousandfold ratio. This is done for clarity of explanation, but in actual design, the resistance values of "two megohms and one megohm" are too large.

The ratio does not need to be a thousandfold, and a smaller value for these two resistances is appropriate. However, I have not supplemented it to that extent in the work.

Furthermore, I wrote that "a capacitor is needed between the AD converter input terminal and ground," but I did not explain why. The reason is that the AD converter input does not have a high impedance, and the signal source impedance must be somewhat low to charge and discharge the sample-and-hold capacitor inside the IC. The reason for adding a capacitor is to ensure that the signal voltage does not change significantly when the sample-and-hold capacitor is charged or discharged.

I was also hesitant about the glitches caused by the ground bounce in the second piece. I thought of a story in which a glitch is caused by something, and the protagonist is ultimately responsible for the cause of the glitch. Still, if I write about a glitch I experienced, it could be considered a leak of secret (know-how). Therefore, the content and cause of this glitch are fictional. I leave it to the reader to decide about the technical reality, but for my part, I have decided to leave it at that.

The protagonist solves the "circuit that uses one AD converter to read two switches" in the first work and the "meaning of sine wave" in the second work in a short time, which is also unlikely in practice. It took me longer. The same is true of the time I learned radio-frequency circuits in the first piece.

Well, it is a novel, so please bear with me. Incidentally, in these two episodes, I intended to depict "thinking with one's head." If you are young

— or even not young — please think with your head.

Another thing I had to wrap my head around was the story. Since it is a novel, I wanted to make it exciting and, if possible, moving. I wanted to portray not only technology but also humanity. Furthermore, I wanted to connect the story of humanity (the protagonist's growth) with the technical story. I leave it to the reader to decide whether I have succeeded in doing so.

In terms of the story, I consciously made it a point to "not create a villain." When there is a villain, we look for the cause of the bad things to that villain. But this is an engineer's novel. In technology, if there is a defect, it is due to design, manufacturing, or usage. In other words, it is caused by humans. We cannot accept the claim that "the laws of physics have a cause."

I intended to reflect this in the story of human nature so that if there is a bad person (the cause of some defect), it is not another person but oneself, the protagonist.

In actual work, there may be "jerks." But even in that case, it is more productive and better for your mental health to think about what you are going to do in light of that fact rather than thinking, "That guy is terrible."

I also mentioned, albeit briefly, the relationship with the customer. Many development engineers may have little direct contact with customers. However, there are always customers beyond the products you develop. This is not limited to the work of engineers, but we should never forget this.

I want electronics engineers to read this novel and feel brighter and more positive. I would be happy if other engineers wrote novels like this one. Also, I would be pleased if middle and high school students read this and thought, "I want to become an electronics engineer." Of course, times differ from when I was a child working with radios and microcomputers. There are more black boxes.

On the other hand, however, we have moved beyond "the age in which an individual can build a computer" to "the age in which an individual can create an FPGA (custom LSI)" and "the age in which an individual can publish a program to the world." Individuals can also create the contents of a CPU in FPGA. And a huge amount of information is available online. We no longer live in a time when people spend all day searching for specialized books at bookstores and end up not finding them, but in a time when online bookstores have too many books they want. I urge young people to be glad they were born at this time.

In principle, this novel's stories, companies, and characters are fictional. There are no models. However, only Manager Yoshioka has a model. He was a friend of mine who passed away from cancer in his forties. This novel is also a requiem for him. I remember once, over drinks, I was saying, "I wish I could write a novel," and he was sitting next to me saying, "Yeah, right." In addition, during the final stages of editing this book, Mr. Ichikawa, the editor-in-chief of ONBOOK, passed away suddenly. I am heartbroken by the death of a man so young, the same age as my friend. Mr. Ichikawa has been a great help to me since my first work, 'Impedance Matching.' This book would not have been born without Mr. Ichikawa. I regret that I cannot express my gratitude in this world. I would like to take this opportunity to express my sincere condolences to both of them.

Finally, I would like to thank everyone who has read this book. Thank you very much.

September, 2010
  Kazuki Sumino

Afterword to Revised Edition

This book was initially published as an on-demand paper book and later as an e-book as an iPad app. However, the iPad app was discontinued as it was no longer compatible due to an iOS update.

Since then, the environment for publishing e-books privately has improved, and the author has now decided to publish directly in e-book(Kindle) format. Accordingly, some contents have been revised.

As noted in the afterword to the first edition, the resistor value is too large in the AD converter episode. Also, the reason why a capacitor is needed was not mentioned in the text. In the revised edition, however, I have included them in the text, albeit very briefly. I also cut the Sekai Camera (iPhone app) scene, which is no longer available. Furthermore, I have stopped quoting lyrics and have eliminated the need to go through and pay JASRAC. And some minor expressions have been changed.

However, the chronological setting of the story has not been changed. The fact that Yuri joined the company just after the ice age of employment or that Yuri's cell phone is not a smartphone has not changed. The same applies to the fact that it is rare for an LSI for the specified low-power radio to have a built-in CPU.

In addition, because on-demand publishing is made-to-order, it inevitably costs more, and the price is higher, but converting to an e-book allowed me to lower the price.

I hope that more people will read this as a result of this.

November, 2018
　　　Kazuki Sumino

Afterword for the English version

Thank you for reading the English version of this novel. In the English version, I have added a few explanations about matters specific to Japan. Other than that, some expressions have been slightly changed in English. However, the rest of the content has not changed.

The author myself translated it into English using DeepL, Google Translation, and Grammarly. I hope you enjoyed it.

December, 2024
Kazuki Sumino

About the Author

Kazuki Sumino was born in Tokyo in 1958 and completed a master's degree in electronic engineering.

Former electronics engineer working for a Japanese manufacturer.

After completing graduate school, joined a Japanese manufacturer. Engaged in developing digital circuits, software, and radio-frequency circuits for embedded devices; left the company in 2016.

The author's books

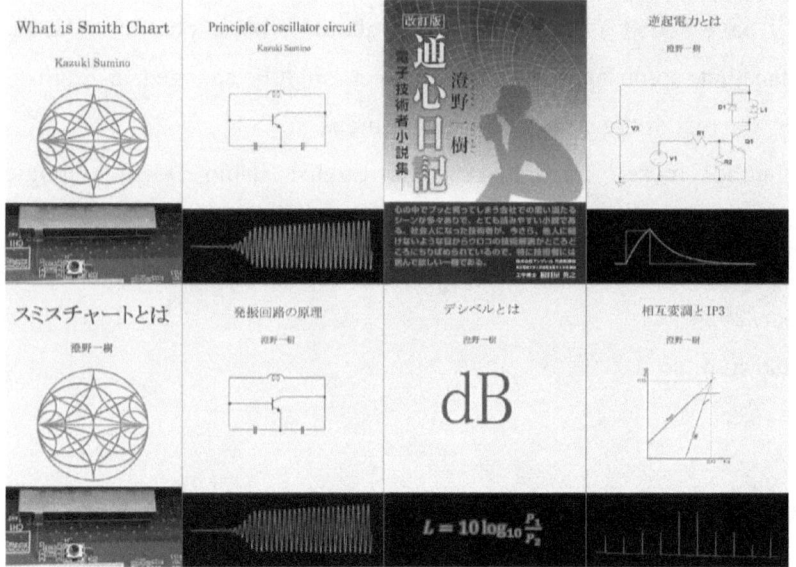

Available at all Amazon stores in the U.S., U.K., Japan, etc.

www.ingramcontent.com/pod-product-compliance
Lightning Source LLC
Chambersburg PA
CBHW031617210526
45464CB00004B/1625